INTO THE
PEATLANDS

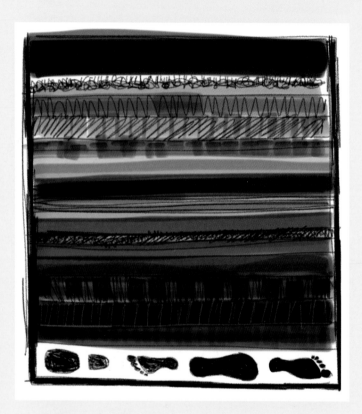

INTO THE
PEATLANDS

A Journey through the Moorland Year

ROBIN A.
CRAWFORD

BIRLINN

Contents

Winter: The Age of Iron

Spring Again: An Age of Folly or a New Golden Age?

List of Plates

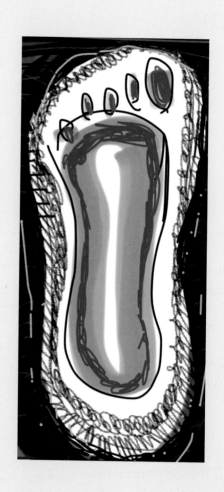

Introduction: A Leaving

Angus Gillies is just one of hundreds of thousands, probably millions, who have emigrated from the peatlands of Scotland. He was born about 1860 on the Hebridean island of Lewis, but like so many before and since he sought a new and better life far away on the other side of the Atlantic. It was an ocean he knew well; twice a day it would either softly wash up the island's beautiful white sandy beaches sparkling luminously turquoise, or crash mercilessly against the ancient rocky cliffs, threatening destruction – sometimes both.

Angus (Aonghas an Gillies in the Scottish Gaelic his people spoke) grew up in Kirkibost, a township of small, self-sufficient farmsteads or crofts on the Atlantic seaboard, and must have gazed over the ocean all his life. People had been living a similar lifestyle on its edge since at least the Iron Age 3,000 years earlier – and probably for longer – growing a few crops, pasturing their livestock on the island's vast moors, taking what they could from the sea and shore. But by the mid-nineteenth century that way of life was under threat as never before and, like so many across the Scottish peatlands, islanders were drawn or forced out to the ever-expanding industrial towns and cities of the mainland, to the central Lowlands, London, the Americas and across the British Empire in the

Opposite: *Peat Footprint.*

hope of a better life. They were victims of a society in flux, experiencing massive social changes following the Industrial Revolution, fleeing poverty, famine, family feud, clearance, dictatorial landlords, oppressive tradition and constricting religion. It was a huge risk to leave for life on the other side of the world, and for some it was fatal, but to stay would have been impossible. So, aged about twenty, he left. Never to return.

More than a century later Marion Laitner, an American, aged ninety, visited Kirkibost. Through family history research she had discovered that she had second, third and fourth cousins living there and as she was reaching the end of her long life she wanted to see the place where her people had originated and the setting for so many of the tales passed down to her. She was welcomed – as so many are – with the generous hospitality of the island and the joy of family reunited, but a further surprise was in store. She was taken out onto the moor where the family had for generations cut the thick, muddy peat to burn as fuel. At a particular spot the top turf was removed for her and there, preserved in the peat, was a footprint – it was the footprint of her father, Angus Gillies.

As she placed her own foot beside the preserved print her cousins explained that on the day before he left his mother had given him a creel and sent Angus out to the bank to collect some peats. Then he was off, like so many of the young islanders. She was desolate at his leaving, but she had to cope the best she could. Life went on.

On her next trip out to the peat bank to collect fuel she discovered one of his footprints there. As the only memento she had of her son she covered the print with turf, and after she died new generations of the family preserved the footprint until that day more than a century later when his daughter returned to her father's homeland.

★   ★   ★

Angus's footprint is not the only one out on the moor. There are prints of other Atlantic emigrants, not always man-made, like those of the white-fronted goose, which leaves the moorlands of the islands for Greenland each spring, or the rasping corncrake, which heads back to Africa in the autumn.

Alongside these are the more recent human footprints preserved at the foot of most worked banks where people in the north and west of Scotland still cut peat for fuel. Among those at the foot of many a peat bank are mine. It has been my passion to try to understand what it is about peat that makes it such a special ingredient in the making of Scotland.

My journey began when I married Angie, who grew up on Lewis. Since then we have returned most years and I have grown fond of the island and its people. I have always been fascinated by built structures in the Scottish landscape – standing stones, subterranean souterrains, cairns, icehouses, water towers, sheep fanks, hydro-electric dams, Roman walls. On Lewis, the peat stacks and cut peat banks are an integral part of the island's culture. Layers of peat slabs, layers of history.

The peat itself is built on its own history, ever changing but still the same. So is the peat stack. It is in a constant state of metamorphosis. From its late summertime construction, slabs are removed daily until it has disappeared, leaving only a dark peaty crumble shadow of its former self. In not a few cases in the peatlands that place has been where the stack has stood for generations, with the people constructing it changing but the family, the home, remaining constant – each daughter and son McLeod, upon NicLeod, upon Mcleod. Come spring and the process of resurrection begins again. In the short months between May and early August the peat is cut, dried, transported and stacked before the long, claustrophobic peatland winter begins.

Peat is a fuel – created by water, dried into a solid, turned into a gas alchemically – but it is also a preserver, an organic time machine. It hasn't only conserved Angus's footprint but also the microscopic pollen grains from millennia ago which were captured in the peat's formation. They can tell us about ancient people's first felling of trees to create agricultural land.

Peat is burnt and turns to ash in the hearth, but miraculously it holds within itself the ashen fallout from Icelandic volcanoes cooked in the belly of the earth, then carried south on the wind; ash from the burning of the forests on Lewis by the Vikings; ash from the peat set alight on Lochar Moss by a spark from a nineteenth-century steam locomotive; ash from Russian peat-fired power stations; or – coming full circle – ash from peat fires raging uncontrollably in Indonesia caused by farmers slashing and burning forest land to feed an ever growing world population.

No less important, it is one of the key ingredients in the making of Scotland's most famous of exports – whisky.

The peatlands themselves are half land, half water. The surface is of living vegetation, but underneath is layer upon layer of dead mosses. One can, with care, walk out into them, but stop and you begin to sink. That has made them places – for humans at least – of seasonal habitation rather than permanent residence. They are transitional places. You journey onto or over them. This book is made up of many of these journeys throughout the year.

The transformations on the peatlands are one of their remarkable characteristics. In the boggy pools, tadpoles turn into frogs; on the heather moors, caterpillars into butterflies; at the shielings, adolescents into adults. To survive on the peatlands, the people have to adapt. Traditionally, crofters are agriculturalists, sheep and cattle farmers, fishers, weavers, sometimes soldiers, singers, storytellers, preachers – or today they are delivery drivers, small business owners, tour guides, vloggers. They have also had to be pastoralists, hunters and gatherers as well.

The moor and peat-cutting have been integral to people's ability to live here. The divide of the year between permanent homestead and temporary summer residence on the moor at the shieling is a link to an ancient transhumance culture that is growing ever weaker, but still exists. One of my aims is to record that ancient way of life whilst the living threads of it still exist.

The peatlands are amazing places rich in wildlife and unique mosses and specialist plants found nowhere else. There is huge concern among naturalists and environmentalists about the future of this special landscape, and the hope is that the moor's ability to capture carbons from the atmosphere offers a partial solution to the centuries of human pollution that have contributed to global warming.

It did, though, become clear to me early on in my journey that of all the amazing creatures that live on or by the peatlands there is one that is almost ignored or dismissed – humans. The people of the peatlands are, like all societies that fringe mainstream global culture, most seriously under threat. After all, my wife and her family, like so many others, left the peatlands and it is highly unlikely that any of them will ever return permanently. As the flow of people drain out of the Highlands and Islands I follow their paths and discover other places where peatlands once existed and find peat-cutting, bogs and mosses still flickering in the fringes of modern Lowland Scotland.

The more I thought about this landscape, the more I envisaged the layers of peat like words and sentences on the page, forming stories of days and weeks; the paragraphs and chapters built up like banks of cut peat running across the moor to months and seasons, until the whole book spread across the full year. This story, going down through the different levels of experience, brought to my mind Ovid's *Metamorphoses*, where the constant transformations of gods and people into other shapes and forms are set within the Ages of Man – from the Golden Age of paradise and innocence to the cruel Age of Iron, war and death. The year I wrote this book, 2017, marked the 2,000th anniversary of Ovid's death. In that time 2 metres of peat have formed. And what of this and coming ages? What will it bring – an Age of Folly or a new Golden Age?

Creation: What Is Peat?

'After leaving the wood the road enters the moor, and is difficult to follow sometimes. The whole aspect of the scene changes. From the corn field and hay meadow you enter at once into a region of moor and peat. You seem to cross the threshold of civilisation, and are transported into a region which bears no impress of the hand of man, and undisturbed by any noisy device or busy handiwork, spreads its fresh beauties before you in all the attraction of nature.'

Peatlands are not dry land. Neither are they wetlands. They are different.

Difference has not always been – and sadly is often still not – appreciated, whether that is a natural habitat, like the bog – the word 'bog' has its origin in the Gaelic word *bogach* – or the people who live on or by it. The idea expressed in the passage above that you 'cross the threshold of civilisation' by entering the moor has led not only to 'uncivilised' behaviour toward this different land but also to the people of that land – 'heathens' supposedly come from the heath.

In early spring I am standing on Kirkconnel Flow in the far south-west of Scotland. Peat-brown water covers the mosses and grasses on which I am walking but, having stopped, it is gradually flooding over my wellies. I am steadily sinking into the ground.

Nearby on 16 November 1771 part of Solway Moss on the Scottish/English border could absorb no more water and 'erupted', or suffered a 'bog burst'. Having reached saturation point, and with virtually nothing tethering it to the land underneath, part of it simply slid away. It became a floating island of peat moving down a gentle slope for a couple of days until it finally formed an adjunct to its old self, which was left between 10 and 30 feet lower than previously.

In *Peat: Its Use and Manufacture* (1907), authors Frederick Gissing and Philip Björling suggest that 'peat' first came into common usage in English following reports of this event, the word previously having been used only in Scotland and northern England, whilst 'turf' was used in the south and Ireland. The *Oxford Dictionary of English Etymology* suggests the origin is Celtic, while the *Collins English Dictionary* suggests: 'C14: Anglo-Latin *peta*, perhaps from Celtic; compare Welsh *peth* thing'.

Whilst the natives of central and southern Scotland and latterly the English simply knew it as 'peat', the Gaels had a whole lexicon for describing its differing natures, textures and uses. It could be a material for burning in the hearth, or a watery mire; sometimes it was the summer pasture of cattle, at other times the pit from which a sinner's soul could only be rescued by the Word of God.

Whatever the name, the same scientific principles for understanding what a bog is apply in every language. Bogs are composed of waterlogged peat created by sphagnum mosses. The structure of these plants acts like a sponge, retaining rainfall and making it difficult for other vegetation to grow, for, unlike other plants, sphagnums need very little nitrogen or minerals to survive. They create a paradise for themselves but one which forms from layer upon layer of decaying mosses. This gradually turns into peat at the very slow rate of about one millimetre per annum. These self-perpetuating conditions are so favourable to the mosses and so unfavourable to normal soil being formed – there are only minute amounts of decaying plant debris for worms, fungi and chemicals to feed on – that it allows for virtually no other plants to grow.

Studies indicate that most bog development began 5,000 to 6,000 years ago, but some are considerably older at 9,000 to 15,000 years. How deep a peat bog is can be extremely variable, with around 1 to 3 metres being the average for blanket bogs, and raised bogs around 5 metres, but it is generally accepted that a bog needs to be a minimum of a half-metre deep. The planning permission for small-scale commercial peat extraction at Tomintoul in the Cairngorms specifies that cutting must stop at this depth to allow the bog to stand a chance of recovery.

Cool, wet and, usually, oceanic climates are ideal for the formation of peat bogs, which explains why in Scotland 13 per cent of the land area (approximately one million hectares) is covered in bog. This is predominantly in the north and west of the country, but there are also significant areas in the eastern uplands too. There are massive peat bogs at this latitude from Russia through to Canada and in northern Europe, particularly around the Baltic, but peat is also found in warmer, wet climates under jungles in Africa and Indonesia.

Raised bogs are predominantly Lowland; blanket bogs, Highland. In a raised bog poor drainage encourages the growth of layer upon layer of sphagnum, which can absorb eight times its own weight in water. As the layers grow, the surface of the bog rises and rises, often forming a dome, swollen up like a pregnant belly. These Lowland bogs suffered in the eighteenth century as agricultural improvement began to be introduced; they were drained to create farmland as the Industrial Revolution saw population growth.

Kirkconnel Flow near Dumfries and Flanders Moss are examples of Lowland raised bogs, while Rannoch Moor in the West Highlands and the Lewis moors in the Outer Hebrides are typical blanket bogs. Local subtleties of geography make for no hard and fast rules for where bogs are created or even the types of bog plants growing in them, but what is clear is that bogs need poorly drained land in which to form, with the amount of water going into the bog exceeding the amount of water escaping. Given Scotland's climate, this can be at high, as well as at low, altitude, with wet enough conditions in troughs high up between peaks being as conducive to sphagnum mosses as they are across the vast flat plain of the Caithness and Sutherland Flow country.

★   ★   ★

Kirkconnel Flow is typical of many raised bogs. It is 98 per cent water and 2 per cent organic matter. The major vegetation is sphagnum moss and cotton grass. Formed 10,000 years ago, it would have grown from pits and depressions left after the last glaciers retreated. As the pits filled with vegetal matter, the sphagnum mosses grew and layer upon layer gradually separated themselves from the land and water around and beneath them, and became almost entirely dependent on rain for nutrients. This type of growth is called 'cloud fed' or ombrotrophic. It is a self-contained hydrological unit, enclosed by hills and outcrops, with very low quantities of nutrients for vegetation to grow except above it. The lagg fen plants and trees growing round the bog's fringes take most of the inflowing nutrients for themselves, so you have this contrast of huge solid trunks of Caledonian pines and silver birches surrounding a moor made up of tiny, delicate waterlogged mosses.

As my wellies sink further into the peaty water, I take a look around me. It's mid-morning and we are suitably kitted out in waterproofs and wellies, heading into Kirkonnel Flow. In the lagg fen birch trees fringe the bog and a willow warbler is invisible but for its song, which starts with a high note then descends as it progresses. Two buzzards wheel overhead, gliding, one in wide circles, the other tighter, but with an amazing gracefulness. A roe deer stands motionless in the woods fringing the bog. Cabbage white butterflies dance drunk on the pale spring sun. In contrast, the talk is of black ravens, appropriate in this transitional place

in graphic form how changes in the vegetation take place over time – different plant groups and different plant species ebbing and flowing. A sample from 6,000 years ago will reveal birch, oak, alder, elm, sedges and some grasses, indicating untouched forest.

Tree pollen travels further than herb pollen, so the sedges indicate bog surrounded by trees with some grasses. Further up the core there is an increase in alder and oak, which means the forest was closing in. The alder likes wet ground, so it was probably right at the edge of the bog, dropping its pollen onto it and causing a decline in the sedges in that area, whilst out on the bog sedges flourished. The trees that like drier ground – oak and elm – were growing on the slopes around the bog.

In the early Neolithic, elm trees across northern Europe suddenly died out. There are many suggested reasons, but elm occupies fertile ground, which is very suitable for farming and, combined with the increase in cereal pollen grains, it may be that the two are linked to Neolithic agriculture.

This seems to suggest that in some cases at least early farmers were the creators of some bogs. The cores show that birch once flourished before the peat was formed and on the surface of the flow they are seeding again. Late twentieth-century government policy was to remove trees from the flow and restrict their growth to the perimeter lagg fen, their thirsty roots devouring the watery soup of the bog, photosynthesising it away into oxygen, drying it out. Nature, though, is no respecter of governments and the top of the flow is again speckled with birch and pine saplings. Some of our fellow corers are quick to pull up these invaders – they are the 'wrong' kind of nature.

'Caledonian pines are not a native species,' rather surprisingly to my ear, exclaims one, an insight into the debates and many attitudes that populate the complex world of conservation.

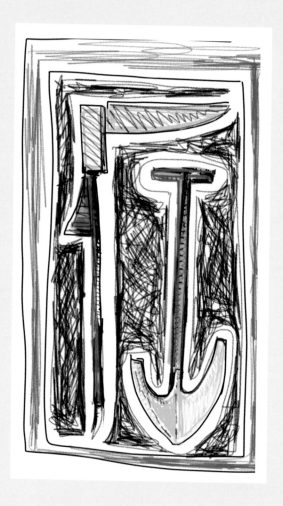

Spring

THE GOLDEN AGE

Paradise Garden

Humans have for millennia imagined, quested for, aspired to, prayed for and attempted to create an ideal world, a heaven on earth, paradise. It might be an oasis in the desert, a temperate valley in the icy mountains or a garden in the city. What are the elements of this desired place? It is unusual, set apart from the everyday; it is enclosed, protected by an outer barrier; special knowledge is needed to enter and navigate; water is key. Different rules apply there. A raised bog may not be a human paradise, but to the plants, animals, birds and insects that live there it is.

Kirkconnel Flow is such a paradise. In dry technical language it is a low-altitude estuarine moss dominated by an abundance of key peat-forming species, but to the senses it is a rich jewel set in the rolling Galloway landscape whose colours and textures transform through the seasons of the year as a precious stone held up to the light. Its domed surface, a swollen belly pregnant with life, is covered with a deep cloth woven by the loom of nature out of multicoloured, multi-textured mosses and sphagnums – Golden Bog-moss ranging from green to chestnut to bright orange, or Austin's Bog-moss hummocked in brown clumps 50 cm tall. Mixed in are *Eriophorum* sedge cotton grasses such as bog cotton, whose fluffy white buds spot and wave in the wind like a cheerleader's pom-

Part title illustration: *Peat Spades*.

poms. There are sweeps of common heather, which brush alongside ericaceous mixtures and some *Molinias* such as purple moor-grass. Other species present, to a greater or lesser degree, are deer grass, cross-leaved heath, liverworts, delicate spring flowers (which turn to fruit on juicy crowberries), glossy bearberries, red cranberries, bog asphodel, Verdigris and yellow lichens, bog rosemary and the less than heavenly (and slightly distastefully named) 'drowned kittens', which grows fluffily in the wetter areas at the edge of the bog. The mosses and liverworts are bryophytes, the most ancient of land plants and the initial colonisers after the ice had retreated. They are so adaptable that they number almost 1,000 sub-species and can be found in many of Scotland's natural habitats, from woodland to mountain extremes, making up a staggering 5 per cent of the planet's total number of bryophytes. They reproduce in two ways: either sexually (very rarely) through spores in a 'sporophyte' stem which disperses through water, or through a cloning-type 'vegetative reproduction' of parent material. Due to their sensitivity to any drying-out of the bog, species like Golden Bog-moss and Austin's Bog-moss are good indicators of a bog's health.

The low nutrient levels force some plants to adapt to conditions using alternative methods to gain vital nitrogen, in the case of Common bladderwort by 'eating' insects. Suited to bogs by having roots which do not need to be attached to soil, it has developed a floating bladder on its leaves which has a flap allowing crustaceans and insects to enter, but never leave, the rotting corpse providing their nutrition. The jewel-like round-leafed sundews are beautiful when discovered but not so for all – perhaps the most well-known insect eater on the bog, it is most similar to the notorious Venus flytrap, with its sticky hairs attracting flies and then gradually closing on the struggling prey. After having extracted all the chemical nutrients it needs, the sundew opens again and the husk of the fly is blown away on the wind. Common butterwort also has sticky hairs at its centre, which trap flies caught in its inward curling leaves. Not so paradisal for all, then.

The sundews flourish because these Lowland bogs are particularly rich in insect life, with many species being discovered there – pulsating orange-bodied damselflies on Bankhead Moss or violet beetles at Black Moss. The pondskaters and waterboatmen of the Red Moss of Balerno row on top of or dive into its peaty pools. These two travellers between worlds are joined in the littoral by others who inhabit both land and water – newts, toads and frogs, whose jellied spawn quivers in the springtime breezes like a quaking mire and magically metamorphoses from egg to tadpole to froglet in the soupy broth of bog. Within the transient world of these mosses other creatures are transformed – squat caterpillars crunching on purple moor-grass turn into peat brown, orange-spotted Scotch Argus or the delicate subtleness of Large Heath butterflies; those on the bog myrtle of the central and western Highlands bloom into the exquisite rareness of the Rannoch Brindled Beauty moth. Larvae become classical-sounding nymphs, then turn into medieval damsel- and dragonflies in the boggy moats to the north of the turf wall built across the slim waist of Caledonia by Roman Emperor Antoninus Pius.

A couplet showing traditional knowledge which has a scientific basis of bog formation begins a letter written in 1865 to the local newspaper protesting about the planned draining of Lochar Moss in Dumfries for agricultural land:

> First a wood, then a sea,
> Now a moss, and aye will be!

As a whilom 'moss-cheeper' – as you smart town gamins were wont to term me . . . you must allow me to enter my protest against this threatened iconoclastic desecration – the reclamation of Lochar Moss . . . In the event of a catastrophe so lamentable as the cultivation of the Moss, what would become of the adders, the wild-ducks, the 'whaups' [Scots: curlew], the stank-hens [Scots: moorhen], the 'Lang-necket herons,' the hares, 'rats and mice, and such small deer,' the indigenous denizens of the Moss?

Look at the valuable additions to our history . . . which are disentombed from the moss every year – flint, Celts, Roman weapons . . .

To him, the Moss is not just a place that connects him to his ancestors, he and the people who live by it are so interwoven with its nature that they themselves are known by the townies of Dumfries as 'moss-cheepers' – meadow pipits and reed buntings – or 'Green Johns'. He continues:

Yet the very mention of Lochar Moss awakes a train of old and dear recollections. Though 'the place which knew us once knows us no longer,' yet it still occupies a niche in the temple of memory and the mind, many a time and oft, goes back to the days 'when my old hat was new' . . .

The moss is a storehouse of memories, a place of innocence where its strange and different nature precluded its intrusion; a place treasured by children for play – where adults, when they intruded, did so briefly – bird hunters and berry gatherers in the youth of the world.

Our Lochar correspondent mentions some of the mammals to be found on the Moss – hares, rats, mice and other 'small deer'. Also to be found grazing its rich surface would be geese and wildfowl, fallow and red deer, wild sheep and goats, and probably cattle too. Remains of all have been found preserved in Scotland's peaty bogs.

As human hunters and gatherers began to turn more and more to settled agriculture in the Neolithic period about 6,000 years ago, the relationship with the bogs, moors and mosses began to change. The move was not a sudden one – the human is a walking creature and, like the seasons, we traverse the globe, moving from place to place, environment to environment, as best meets our needs at the time.

Today we are still restless creatures, our footprints measured in carbon.

The Lewis Moor

In the oceanic climate of north-western Europe the seasons may be wet, but except at high altitude they are relatively mild – if infrequently hot. This dampness, which creates the bogs and allows them to flourish, discourages settlement and allows for habitation only on the fringes, or as a summer residence. These moors are almost all blanket bogs, a different environment from the Lowland raised bogs and mosses like Kirkconnel Flow. The one I know best is the moor that covers the north tip of the island of Lewis, itself the most northerly of the Western Isles.

The quiet and calm of air travel allows for a different view; height and distance give a wider perspective. At 36,000 feet over the North Lewis moor on a flight from Reykjavik to Edinburgh the only suggestion of movement in this vast panorama comes from waves breaking on a beach. There is no distracting detail, no hectoring by plover, lapwing, skua or biting wind, or (very occasional) overheating sun from above. Nor is there the shock of walking on this moor only to find myself suddenly sinking, then frantically stumbling to set foot on solid ground again, with the realisation that I have five hours' more walking to do with wet, boggy feet. No, from above all is calm, dry, gently throbbing, unworldly. The relentless waves thud and crumple silently on the shore below, line after line creating grain after grain of sand; they are the shuttle, the strands of seaweed, the threads on a silent weaver's loom.

Other lines are being laid down, imperceptibly. From a height of years you could look back and see the peat forming, millimetre by millimetre, year by year. Like flotsam washed up on the shore and gradually covered by sand, the sphagnum takes into itself the history of the atoms that wash up on its surface in all their myriad molecular diversity. Gradations of years see news become history, history turn into ancient history, then pre-history in its peaty strata: last summer's lost tweed cap; nuclear fallout from Chernobyl; a drained beer bottle thrust into the wet peat bank at the end of a day's cutting; the carcass of a lost sheep; a Bronze Age axe used to fell the trees that allowed the peat to originally form. But this strata is a mere fringe on the cloth that covers this earth. The rock below has a deeper strata: pink Torridonian sandstone blushes at its own youth compared to the ancient Lewisian gneiss, the bedrock of this island that occasionally breaks through the moor or is exposed by cutters under the peat. Yet back on Iceland even the Torridonian rock would seem old, as the continental plates of Europe and America grind apart and cast up, in quaking volcanic cataclysms, land younger than this peat.

For now, all remains cartographically still – but for that white line of waves breaking along the mile and a half of sandy beach at Tolsta. It is, after all, the Sabbath.

★   ★   ★

Under the peat, the bituminous oils are moving. Having retreated before the winter's chill, they have joined that secret underground blossoming, invisible beneath the unchanging sphagnum, the season-less lichens, that heralds spring. We are on the threshold of the seasons.

Glacial coldness still whips across the North Atlantic. Inside the traditional *taigh dubh*, the island's blackhouses, where humans and cattle used to share their winters on Lewis, both would be sensing the change – the last two weeks of winter and the first two weeks of spring, *faoilteach*, are welcomed hospitably.

The beasts would be straining against the *bacan*, the tether stake built into their stall. As winter progressed, the point to which they were tethered had to be raised as, like some kind of super-speed peat, the strata of their dung built up the floor level higher and higher. All inside are eager to be out: the crofter to be digging up the dung and spreading it on the land to fertilise the poor soil before sowing this year's crop; the women and children anticipating the happier, more relaxed mainly man-free days of summer pasturing on the moor; the starving beasts slavering for tender new shoots; the youths looking forward to courting and flirtations at the shielings.

For the island culture was until recently a life divided between a winter and a summer home. The ancient transhumance way of life was lived – part agriculturist, part pastoralist, part hunter-gatherer. The contribution of all was needed in a society based in small townships consisting of family units which subsistence farmed and relied on communal assistance. Their lives were lived half the year in smallholdings or crofts, dividing their time between small-scale arable agriculture, growing cereal crops (oats, some wheat, 'clover' grass for cattle feed) and vegetables (carrots and turnips, and later potatoes), or raising a few cattle, goats and sheep, with perhaps a rare horse – the Lewis pony (a breed now extinct) – and some poultry, a little inshore fishing and periods spent working in obligation – obligation to the tacksmen who sub-rented land to the crofter from the landowner, who was often their kinsman or clan chief.

In some times past this obligation would take the form of fighting for the clan chief or for those to whom he owed obligation, such as the Lord of the Isles or the King – or would-be king – of Scots. Obligated but also independent, or as Jessie Kesson describes crofters and their community:

> . . . bold whitewashed houses, their windows searching the ocean,
> like their inhabitants, with an eye always on the sea; being dependent

on both the land and the fishing had rendered them curiously independent of either . . . crofts forced themselves up out of the earth in small defiant protest.

The land behind each croft was both limited and lacking in nutrients, hence the need for early fertilising with dung and then adding further nutrients throughout the growing season – seaweed collected from the shore and burnt, the old peat smoke-tarred roof straw: the completeness of peatland life, everything interwoven, utilised to ensure survival.

With prime land given over to agriculture, cattle and sheep would be grazed on the poorer land, which was shared communally – in Ness that land was, and still is, the moor. The crofters would lead the cattle out there each spring and return in the autumn, usually under the care of young women. This could be a happy time.

Whilst on the moor pastures, they would stay in shielings, the family's summer dwelling, a more basic form of house – huts constructed or reconstructed annually in the same location of stone foundations, turf walls and perhaps a timbered roof covered with turfs. Doors and roof timbers would be brought out and returned to the winter croft at the beginning and end of the season. Some circular rubble-built shielings modelled on an ancient beehive construction would also be used where stones were available. Each of the one-room shielings would have a fireplace, a bed space and niches or 'cupboards' built into the walls for storing milk, cream, 'crowdie' cheese and other dairy products made from the cows' milk. These stone 'presses' are of a design built into the walls of now-abandoned crofts on the 'cleared' lands of Caithness on the mainland, as well as twentieth-century tenements in Glasgow, Edinburgh and Dundee, or the 5,000-year-old walls of the Neolithic settlement at Skara Brae on Orkney. Sometimes after milking, the milk would be taken back to the croft in the evenings by boys and girls, latterly on bicycles. In some shielings there would be two doors, used alternately depending on the direction of the wind. If there was a window, it would often just be an opening in the roof, covered with a turf at night or when it rained. Women would also spend time at the shieling collecting lichens and mosses from the moor for dying wool, which would then be woven into the wonderful tweed material unique to the Western Isles, or they would collect plants for herbal remedies or cooking. After a hard day's work on the croft, fishing or away on some paid work, or work of obligation, men too would sometimes come out to the shieling, perhaps bringing some lush weeds cleared from the field which the girls would then give to the cows as they were milked to soothe them and sweeten the milk.

Alexander Carmichael, the great nineteenth-century recorder of traditional Gaelic life, prayers and blessings, records:

> Most of the shielings were several miles, some six or eight, some twelve or fourteen miles, from the townland homes. The moorlands are rough and rugged and full of swamps and channels, and the people use much care in guiding the cattle ... It is instructive to see the caution with which the older animals travel over the rough channelled moors, daintily feeling their way when not sure of their ground.

In his notes, Carmichael tells us that the herder would sing to the cattle as they were driven and they would keep time with the music, Highland cattle being well known for their intelligence.

On bringing the cattle home, a young animal may have separated itself from the herd:

> The herdsman, fearing that the truant may have been caught in a bog or fallen over a rock, searches high and low, near and far. At last coming in sight of him, he addresses him in terms and tones different from those he used to the others. The animal looks up;

it is only for a moment: he is off at his hardest, taking the nearest way for home ...'*Tuigidh an cu a choire fein* (The dog understands his own fault).'

Carmichael further warns against the dangers of the cut peat bank:

Charm for Stock

The charm placed of Brigit
About her neat, about her kine,
About her horses, about her goats,
About her sheep, about her lambs:

Each day and each night
In heat and cold
Each early and late
In darkness and light;

To keep them from marsh,
To keep them from rock,
To keep them from pit,
To keep them from bank ...

When I send a photo of Lewis taken from the Reykjavik plane to my friend Mary showing her croft from above, she tells me that her husband, John, has been out on the moor turfing two peat banks in preparation. 'Turfing' involves removing the top living 6 inches of the peat bank at about shoulder height and replacing it on the moor to regrow at foot level. The exposed peat is then ready to be cut. In previous years, this was the time the shieling would be repaired or rebuilt in preparation.

All is anticipation.

First Experience

Fifteen years before my flight from Iceland and 36,000 feet lower, Angie and I were on the beach at Tolsta. We were about to embark on what was then termed the 'Western Isles Walk'. It was, and is, intended to be a footpath. It does not, as the name suggests, encompass a trek up or down the length of the islands that make up the archipelago, let alone weft and weave or eightsome reel in and out of the Inner and Outer Hebrides. It does not, like the West Highland Way, draw youthful, be-kilted, urbanite, Duke of Edinburgh Award hikers over the course of four or five days to offer their peely-wally knees to the massed ranks of the Loch Lomond and Rannoch Moor midges. It is not the middle-aged whisky enthusiast's elongated pub/distillery reel/crawl of the Speyside Way. Nor is it the cyclist's Loch Ness monster-spotting tour of the Great Glen Way. No, the Western Isles Walk is a ramble up the bleak, brown east coast of Lewis, linking the green patches of land at Tolsta to those at the northern tip of the island at Ness.

The route winds its long and desolate way over the North Lewis moor and, though a recognised walking route, the reality suggests that one day, many years ago, someone with a quad bike drove along, banging in some marker posts every mile, but so few people have travelled along it there is no discernible path. 'Follow the clearly sited green and yellow waymarkers, guiding you across the moor to Nis' instructs the sign at the head of the beautiful sands of Traigh Ghearadha, but ominously the small print states, 'The Western Isle Tourist Board is not legally responsible for any accident or injury to persons undertaking this walk.'

The starting point is a warning – the 'Bridge to Nowhere', the remnant of a failed 1920s attempt to join the north of the island to the main population centre on the east coast at Stornoway. There is still no direct route. As recently as the 1960s, Mary from her home at Skigersta in Ness had to get up at 5 a.m. on a Monday morning to get to the Nicolson Institute, the only secondary school on the island. She would walk the 3 miles to the main road, whatever the weather, and catch the 'workers' bus' to Stornoway, 26 miles away. Arriving there, she would go to a café for breakfast before starting her school week, which involved staying in the school hostel, like so many children from all over Lewis and the conjoined island of Harris, because transport links were so bad. Sometimes she would return home for the weekend; sometimes she would have to stay for a fortnight.

Today, Mary's journey takes forty-two minutes by car. These journeys are, of course, modern routes. Until the fairly recent past the majority of inhabitants of Ness would not need or want to travel all the way to the town of Stornoway, which itself has only grown to any size in the past 200 years. The need for education to secondary level, let alone the education of a girl, was unthought-of except in the rarest of cases. Lives were lived within a far shorter radius of the home; everyone was needed on the croft to make life a success.

As we pass the Bridge to Nowhere and continue onto the bleak moor, it is clear that we are walking into a dangerous place. The water/land proportions vary across this 10-mile stretch of the huge expanse of blanket bog, the second largest in Scotland and one of the largest peatlands in Europe. It makes for a splashy and soggy walk, as one minute you are on firm ground, the next you are not.

According to DEFRA, the Lewis moor is 'probably the most extremely

"Atlantic" of all the blanket mires in the UK and indeed Europe' and in scientific terms, the vegetation is

> predominantly, though not exclusively, of the *M17 Scirpus cespitosus – Eriophorum vaginatum* blanket mire type, with purple moor-grass *Molinia caerulea* and deer grass *Trichophorum cespitosum* often dominant and accompanied by cross-leaved heath *Erica tetralix*, bell heather *E. cinerea* and the western bryophytes *Campylopus atrovirens* and *Pleurozia purpurea*. One particularly characteristic feature is the widespread occurrence of the woolly fringe-moss *Racomitrium lanuginosum*. Although this species is quite common as a hummock-former in northern and western blanket bogs, particularly in areas of peat erosion where it caps the remaining peat haggs, it is only in the extreme north-west that it forms extensive carpets, a niche which elsewhere is the preserve of bog-mosses *Sphagnum spp*. A mosaic of bog habitats is present including pools, depressions and small lochans with wet and dry heath on intervening knolls. A number of oligotrophic lochs lie within the site as well as the headwaters of numerous small rivers and streams.

This description uses a scientific language, with its Latin that is both universally understandable to, say, biologists and botanists across the globe, but particularly difficult to comprehend for the majority. In English there are many bog-words, but not enough to describe this moor as succinctly or accurately as in Gaelic. This may change in the future, as languages merge and grow, but at present it is not possible to sum up the nature of this land in one single word, while in the mouths of the people of Ness there is a whole lexicon of moor words. Take, for example, *botann*. If you graze your cattle on the moor, you need a word for 'a hole, often wet, where an animal might get stuck'. In Edinburgh, London and New York you do not need a word like *breunlach*, never encountering there 'a sinking bog. May be bright green and grassy, or open water with vertical sides and a peaty bottom. Dangerous to step into.' Even a description that needs nine words in English can be communicated in a Gaelic two-word compound, *lèig-chruthaich* – 'quivering bog with water trapped underground and intact surface'.

Then there are the water-based words for this water-based landscape, some corresponding with English: *allt* – stream; *abhainn* and *glean* – rivers; *clachan sinteag* – stepping stones. Others are descriptive terms unique to this land. Anne Campbell's *Rathad an Isein: The Bird's Road* lists: *go* – 'a gravelly river', *feadan* – 'a stream running from a moorland loch; smaller than an *allt*', and *caochan* – 'a stream which is so obscured by vegetation that it is (virtually) hidden'.

The moor to the north of Tolsta, never far from the sea, has myriad *bhat* (lochs and lochans), in places making up a quarter of its surface. Further west, the open water becomes even more abundant, concentrated within multiple tiny lochans, pools and ponds which freckle its face on the Blar nam Faoileag, the bog of the white gull, or of the white crest of a wave.

Like sailors reading the sea (which these island crofters also were), they knew and understood the land so alien to outsiders. They knew, had to know, how to navigate this land, a dangerous place to the uninitiated and the innocent.

On a gravestone:

Ealasaid (Elizabeth), 6, drowned on the moor.

Lewis peat bogs are deep. Underfoot the peat feels different. It is muddy, squelchy, sucking at your boot, trying to grasp you, draw you in. False solidity gradually liquefies and using your own bodyweight sinks you deeper into itself, smothering until you are overwhelmed, suffocated. Siren.

Not only are you at risk from the moor but also its inhabitants – not as you may think from the gentle, subtle adder, the poisonous snake of the bog, or even the carnivorous sundew plants, but, as I found out, from aerial assault from the skuas nesting along the moor's sea-cliff fringe, which will dive-bomb you if you stray too close to their eggs or young, squirting you with a fishy excrement of the foulest nature. Whilst not as threatening as the skuas, other staunch defenders of their territory are the stout-hearted golden plover, or the lapwing vigorously protecting its nest with a vocal flourish.

Then there is the supernatural danger, for in loch and lochan out on the moor lives the *each-uisge*, the water horse, or in Scots, the kelpie.

This is a traditional story.

Once the cold winter winds had died down and the flowers began to bloom a crofter called Domnhall (Donald) went out on the moor to re-open his summer shieling on the hillside above a big loch. Although the land provided rich pasture for his cattle to produce the sweetest milk and plenty of grazing to fatten up his sheep and lambs, his neighbours thought him mad. None would venture out to that part of the moor after dark, for they believed that in the black, mysterious water lived a fearsome monster, a water horse.

Tales were told of many a man last spotted heading for the area, never to be seen again. The *larach airigh* (remains of old shielings) were the only signs left of those eaten by the monster over the years. And how Domnhall could let his beautiful daughter, Peigi, (Margaret, or Peggy) stay in that place they had no idea, for there were places out on the moors called *àirigh na h-aon oidhche* (the shieling of the one night), where an *each-uisge* had feasted on a young woman in the past.

'Build on the other side of the *allt* that flows down into the loch,' they would advise. 'The kelpie cannot cross flowing water.' But he would not listen to them.

Peigi was fearful and wary each evening when she went down to the loch-side to bring the cattle back to the shieling for the night. She knew that the stone walls would stand strong but that the turf door and roof would crumble if attacked. In the morning her fears would pass and she would spend the day collecting mosses, lichens and crottal from the rocks of the moor to dye her wool beautiful colours for weaving into tweed during the long winter months.

One morning as she sat by the door spinning her wool a dark shadow passed over her. Looking up, she saw a young man soaking wet from head to foot.

'Sorry, I did not mean to frighten you,' he said.

'How is it that you are so wet?' she asked.

'I was caught in a sudden downpour as I was coming round the other side of the hill,' he replied. 'I'll soon be dried by the sun and wind.'

He sat beside her and as this handsome stranger chatted away to her a pleasant, drowsy feeling started to creep over her like a dream. He asked if he could lay his head on her lap and if she would comb his wet hair. Peigi could not help being alarmed; there was something strange about him, despite the pleasant feeling. As she teased it out, she noticed that in amongst his wet hair there was green weed and gravel – just like the gravel and weed at the bottom of her father's nets when he fished in the loch. She struggled to shake off the torpor which she quickly realised was being generated by this strange man, if man he was. For was he not the *each-uisge* itself?

Quick as a flash Peigi was on her feet, tearing across the hillside with the water horse no longer in the shape of a man but in all its fearsome glory, hot on her heels. It was perfectly adapted for the watery moor, with the front hooves of a horse and the tail of a fish behind. Had Peigi not been spinning and quick-thinking, it would surely have caught and taken her back to its lair deep under the loch, but as she fled she cast the wool from her hand and it entangled itself around the beast, slowing it down and giving her just enough time to reach and cross the flowing burn and so to safety.

After hearing her tale Domnhall regretted not listening to his neighbours and was very grateful to still have his wise, brave and strong daughter. From then on, he built his shieling beside those of the other villagers. If you know where to look, you can just make out the *larach* of Peigi's shieling on the hillside, but I don't recommend you go there.

But to the brave, strong and cocky of legend even an *each-uisge* holds no fear. In the traditional tale 'Big MacVurich and the Monster', the hero returning from the hunt with a stag on his back finds a tiny little creature at the edge of the loch, which he picks up and puts in his warm pocket. Hardly is he home before a monster appears outside his house demanding her whelp's return.

'For a bargain,' replies Big MacVurich. 'Build me a causeway across Loch Dubh on which my peats may be brought home . . .'

'Alas, alas,' says the monster, 'though hard the saying, it is better to fulfil it.'

In the dead of night, the monster comes and calls out at the window: 'That is ready, Big MacVurich. Out with my bouncing boy.'

'No, except for a bargain,' says Big MacVurich.

'Let me hear thy terms, thou wicked man,' replies the monster.

'That thou bring home every single peat I have on the hill slope and make a stack of them on the hill at the end of the house.'

After completing this task and building Big MacVurich a stack the monster finally gets her child back.

MacVurich is not the only man out on the moor. Halfway between reality and story lurks Mac an t-Sronaich. He is a mysterious bogey-man who was outlawed on the wild Lewis Moors. Little is known about him. Some say he was a serial killer called Alexander Stronach from near Garve on the mainland who took refuge on the moors of Lewis in the early nineteenth century. It is said that for years he preyed on people, sheep and cattle, murdering and killing, becoming a figure of fear. Yet no court records exist to confirm any stories. Real or imagined, the threat of Mac an t-Sronaich has haunted many a child and adult over the years. Countless tales about him have been passed down through the generations. There is hardly a cave or ruined shieling on the moors from Uig to Ness and Tolsta which is not said to have housed him. One such cave where he was supposed to have hidden can even today inspire fear – I just about jumped out of my skin when the bulging devil-eyes of a black-faced sheep suddenly appeared out of the darkness.

It is said that Mac an t-Sronaich was eventually captured and taken to Inverness for trial and after the passing of the death sentence was asked if he had any regrets, to which he replied: 'I regret drowning a child and not murdering a minister.' These are the kind of stories you would hear told by night-time fishermen out on the moor. Traditional knowledge, as well as stories, is passed down through the generations. White pebbles of ancient Lewisian gneiss rock glinting with mica are dropped into lochans, reflecting the moonlight and attracting salmon.

★ ★ ★

From the lochs and lochans, the brown peat burns remorselessly try to find their way to the sea. Across an invisible-to-fathom watershed some flow east to the Minch, others though rising further east flow west to the Atlantic. Sometimes they trickle invisibly under the skin of the peatlands like the veins and arteries of the bog's body, a hollow gurgling sound all there is to alert you to their quaggy presence.

In this tree-less landscape the river is the brown trunk of a water system that branches for miles back into the peatlands. Sitting over their tea and scones, the island hosts can, bard-like, recount from memory the generations that spread back into the past, linking their visiting Canadian cousins eight times removed to this place.

Spreading like the roots of the moor's family tree, they tell its story through the ages – the successful branches lush with life, others end up in dry and barren oxidised peaty cracklings.

The coastline eventually slices like a *tairsgeir* through the moor and the waters gush down the cliffs and white sand to the sea, staining the blue-green waves a milky-tea colour. Like oil, the brackish moor water floats atop the salty waters of the Minch.

Waves are constant in their thudding into the Lewisian gneiss and will eventually wear it down. On the sandy remains of this grinding, little piping birds and the occasional giggling grandmother – neither wanting to get their feet wet – dance up and down with the ebb and flow of the tide. Further along the shoreline the Atlantic is eating into the softer strata of the land, nibbling at its base, gradually eroding it.

Looking up from sea level, I am confronted by a miniature peat-coloured cliff 15 to 20 feet high. In some places the soil has collapsed onto the rocks below, in others the land overhangs and is drooping disconcertingly. You find yourself looking up, underneath a herd of cows grazing the machair, precariously close to the edge.

As I am thinking of the beer-coloured waves, of the slaked thirst of cutters, my brother sends me a photo of the beer he is about to open, a Loch Lomond Brewery 'Peat Smoked Ale'.

'What does it taste like?' I text back.

'Smells like an Islay whisky with a slightly medicinal aroma, first a malty sweet biscuit taste – Tunnock's caramel – then the big peat hit, a sooty chimney!'

At 5.4% it is a strong or 'heavy' Scots ale. In America there have been similar 'wee heavy' modern beers crafted using peated malt. Beer has a long history in the peatlands, dating back millennia. Heather ale – *fraoch* in Gaelic – is still brewed commercially using heather and bog myrtle among its ingredients. In the late eighteenth century, Thomas Pennant was making a tour of the Hebrides and wrote, 'Ale is frequently made of the young tops of heath, mixing two-thirds of that plant with one of malt, sometimes adding hops.'

Ping!

Next up from Loch Lomond Brewery, a 'Silkie Stout'.

★　　★　　★

Remnants of the past litter this land. Ridges and furrows of 'lazy beds', where peat was cut and seaweed, washed of its salt, was piled onto rows of potatoes, rib the sandy, fertile land near the sea. Rising up by the cliffs is Dùn Othail, the stronghold of the illegitimate son of the last Chief of the McLeods of Lewis, who was executed in 1597. This fortress sits atop a sea stack, and at the end of the Western Isles Walk is another – Dùn Èistean – in the territory of the McLeods' bitter enemies, the Morrisons of Ness.

Above the beautiful beach at Eoropie, at the Butt of Lewis, the Morrisons still proudly show the Clach na Fala, a stone with red-orange colour said to be stained by the blood of their enemies' heads being smashed against it.

A tragic, ruined house at Cuilatotar tells the sorry tale of Iain Buidhe (John MacDonald) of Ness. Press-ganged during the Napoleonic Wars into the Seaforth Highlanders, he served all over the British Empire only to return in 1820 after many years to find his family evicted and exiled. He saw out his lonely days in this desolate spot.

Along the route of the walk a beehive-shaped shieling of very ancient design is broken down but still in a repairable condition. Other shielings further on at Cuidhsiadar are still used – treasured family retreats, some of stone but most of modern materials – wood, corrugated metal – all beautifully maintained. Some families have a bank of peat near their shieling from which they cut fuel for their own use. These are true summer houses, with a beautiful sandy beach, loch fishing and moorland activities of all sorts. A small chapel was built here for the worship of God by people staying at the shielings. More modern constructions –

radio masts (and between my first visit and now a mobile phone mast too), wind turbines and a water tower and lighthouse – signal the coming of permanent habitation towards the end of a long walk. After many toilsome, foot-sucking miles, definable tracks, then paths, then peat roads lead gradually off the moor and into the community of Ness.

The Minch on our right all the way up the coast now gives way to the Butt of Lewis, the famous lighthouse marking the most northerly point of the Hebrides. On the left, the huge Atlantic – an ocean not a sea, and nothing between here and America. Jacob's ladders of sunlight shaft down from grey skies to grey ocean momentarily lighting small patches of its surface, emphasising its hugeness. Ness is the Norse for a headland – almost all the names on Lewis are of Viking rather than Celtic origin; Còig Peighinnean Nis is so-called because it is made up of five pennylands, a Norse measurement of land. The Vikings are believed to have avoided the dangerous passage round the northern cliffs of the Butt by hauling their ships out of the water and across the narrow strip of land between Eoropie on the west coast and Port Ness on the east. It is said that the Norsemen's main contribution to the moors of Lewis was to burn all the trees in the island in a fire that could be seen for days from the mainland.

From the ashes of these fires island people rebuilt their lives, adapting to the rages of weather and fate, and the whims of those with power over them down through the centuries. These particular circumstances have shaped the people and their environment. From the crest of the last rolling moorland hill, the idiosyncratic pattern of the farmsteads is laid out before us.

Though half a mile long, croftlands can be only 15 metres wide. Some are grazed with sheep, some filled with crops, many with just grass. Like agriculture, gardening is tough here, the wind the dominant factor. Some houses have built up a barrier of small hardy shrubs rising to hedges, then small trees to protect against the wind, and in these enclosed paradises something like a normal garden can flourish – little did I know then that a decade later I would be chatting about Robinson Crusoe with the inhabitant of one of these moorland desert islands.

There is no shop here at the very northern tip of the island, though the Stag Bakery van from Stornoway visits on a Thursday, a vital lifeline to those without transport.

My mother-in-law still tells (with a little pathos) how, in the 1980s, with a husband building houses in the south of England and four children under eight years old, the highlight of her day would be to go out to the baker's van – usually followed by a procession made up of children, a couple of lambs, a line of chickens and the cockerel running back and forth with the youngest child's dummy in its mouth.

The bakery has a fine Scots duality: couthy local bakery on the island, and on the mainland I recently bought a packet of Stag Bakery seaweed water biscuits in a very upmarket city delicatessen: 'This deliciously tasty cocktail size biscuit evokes memories of the coastal holiday, of islands and a brisk sea breeze.' We laughed at the description of the 'brisk' Atlantic 'breeze'!

With an Atlantic 'breeze' gusting at the end of our walk, no Stag Bakery van appears but magically, in a truly Western Isles way, in a driveway beside the community centre, a fish and chip van. We join the queue behind a black cat waiting, hopefully, to be served. One of the best meals ever.

It fortifies us for the long wait for a bus.

In the thirteenth-century St Moluag's Church (Teampull Mholuaidh) – Episcopalian not Presbyterian – a prayer is written on a wooden clothes peg. In answer to ours, the coach back to Stornoway appears from round the side of a croft.

★　　★　　★

The following year we were back in Lewis, this time cycling along the west coast road. When Angie saw me fall, she thought I'd broken my back. It wasn't just that the road ran on a causeway above the moor but that I'd fallen across the drainage ditch and my spine appeared to bend backwards as I landed. Luckily I had a rucksack on, which bore the impact, and the bike had continued down the road on its own for a few metres, so the initial 'thud', though winding, was relatively OK. What was worse was my then slow but steady descent into the black peaty depths of the ditch.

First was the shock of hitting the water. Sudden immersion, even on a summer's day, isn't enjoyable. And 'summer' is relative; we're talking Lewis in early August – 12 degrees, raining and a medium southwesterly headwind driving the rain firmly into our faces. The only advantage being that I was already in full waterproofs. Despite that, I was soaked, the mirey water finding its way into my clothing. As it runkled up, my head submerged.

Fortunately, Angie was pulling me out in seconds and all was well. We were soon on our way again, and it wasn't long before we were home and I was discovering, under the shower, how many moorland creatures you can accommodate about your intimate person after a relatively brief visit to their environment.

The cause of my fall was over-enthusiastic excitement at seeing a buzzard rise and take off from a fence post as we cycled past. Nowadays, dulled by country living, I would not have suffered the same fate, being well accustomed to daily encounters with these lazy 'tourist eagles', just as my son fails to be excited by tractors, but back then we were city boys and medium-sized raptors and red peat tractors were high in the 'get excited and point' stakes.

Despite the appearance of bleakness and emptiness, the moor, as my brief encounter revealed, is rich in a whole range of flora and fauna . . .

Littoral: Beltane

At Callanish, the magnificent standing stones were almost entirely covered by blanket bog to a depth of 1.5 metres over the course of 2,800 years until excavated in 1857. The ancient mysteries and rites practised there have not been preserved in the strata of the peat, and though it is mainly just the sound of the ever-present wind that briskly catches your ear there today, echoes of stories of battles between eagles and men petrified occasionally sound faintly down the centuries.

Alexander Carmichael was fascinated by the ritual prayers, songs and traditions of the Highlands and Islands, and in the material he collected in the mainly Catholic southern islands of the Outer Hebrides he found traces of pre-Christian rites, customs and beliefs.

The northernmost islands, being strictly Presbyterian, have tried to erase all traces of this religious culture in the purity of 'the Word' but so enmeshed are they in the culture that they can still be found, especially in the practical everyday – or every season – tasks that have not changed since people lived and worshipped by and on the moor.

Fire remains part of the peatland culture.

An ancient Beltane ritual is celebrated on the first of May, when cattle and people pass through fire from winter croft to summer pasture on

Opposite: *Aithinne – Peat Torch.*

their way to the moorland shielings. Other practices to do with the cleansing properties of fire are associated with this time of year. Passing fire round an infant thrice is a tradition that is supposed to prevent the baby being swapped by the fairy folk for a changeling. In *The Silver Bough*, F. Marian McNeill's classic work on Scottish folklore and folk belief, she writes:

> The Quarter Days being holy days, occult influences were believed to be more potent and magical rites more effective than at other times. Hence the ritual kindling of the need-fire, the supreme protective against disease, disaster, and the powers of evil; the saining [blessing] of cattle and crops, boats and buildings; the visits of the sick, the maimed and the barren to the holy wells; the divination of rites; and the communion of baking and dedication of the sacrificial cakes. They were also lucky days for setting out on a journey, for a new undertaking, and for drawing lovers together.

Christian prayers and the singing of psalms even among very small groups of people start and finish many daily and seasonal tasks in these peatlands today. Prayers, and a long look at the BBC weather forecast, are the precursors of many a day scheduled for cutting the peat.

black parallel lines (sometimes dotted) which means 'other road, drive or track, fenced and unfenced'.

If you know these routes because they have, for generations, been part of your family's history, there's no need for a map showing *rathad nam banachagan* – 'the road of the cattle-herds/milkers', the route to the shieling. Animals may not read maps, but a sheep will have *innis* and *astar* – a natural instinct to return to the area of moor where they spent their first summer. In the burns, the salmon run unguided, back from across the Atlantic; overhead, the returning African swallows. On the bog it's slightly more confusing, especially when the weather sets in, or at night.

For seeing in the dark, people would use a peat torch – *aithinne* in Gaelic – which gave a good light and could be used for showing the way, or to fish for salmon and trout out on the moor at night. The human ability to make fire confuses other creatures. Later that evening, under the clear electric light, the night outside a simple black, I mark my travels across the moor on a map and empathise with a moth's bewildered journeying across the window pane. But for humans a lack of sunlight aids confusion; when the fog sets in, especially common in these wet lands, anyone can get disorientated.

The classic staple of fairy and cautionary tales is a child lost in the wild. Mrs Campbell of Barra remembered this story told to her in her youth about a girl left out on the moor:

As a child she often used to go out to the moor for peats. With the peats being heavy and her being wee, she had to rest frequently on the way home. On one occasion fog came over while she was resting and she lost her way. She was found by an old couple, who emptied the peat from her bag then abandoned her. After wandering around for many hours she eventually found her own way home.

Who were the old couple? Why did they empty her bag? Would they not help a child home?

In 1867, four-year-old Alasdair MacDonald was not so lucky. Deciding to follow his mother with her creel on her back out to the peats, he lost his way and drowned in Loch Shiabhat.

★   ★   ★

The cutters have left their tools on the moor – *tairsgeir* and spade – the blade of the *tairsgeir* stuck into the wet peat at the foot of the bank. The derivation is 'turf' and '*sgian*' – blade for cutting turfs. Different North Atlantic cultures or even islands had slightly different methods of cutting and tools – in Shetland the 'turfing' would be done with a long-bladed 'ripper', not too dissimilar from an Icelandic turf knife, *torfljar*. Indeed, the Shetland term for this is 'flaa-ing (flaying) the bank'.

I pull the *tairsgeir* out for a look, the blade probably made by Calum MacLeod or previously by his father in their blacksmith's forge in Stornoway. He reported an upsurge in demand after recent oil price rises. The wooden shaft is remarkably light, almost like the dead wood found in the bog. Dry with no oils, it is testament to the softness of the peat that no heavy duty shaft is required to drive the *sgian* into it. A cutter might soak the *tairsgeir* for three weeks before cutting, swelling the wood into the *na h-eilean*, the metal socket above the blade. I carefully and respectfully replace it – later we shall hear of the dangers from the Fairy Folk of leaving peat spades on the moor.

The peat-cutting crew is the *sgiobadh*; the cutter, *buain/leagail*; the thrower, *tilgeil/saldail*. The blue boilersuit – often with Western Isles ferry operator Caledonian MacBrayne's logo – is the workwear of choice for many cutters.

So, how is peat actually cut?

Peat-cutting Day:
Latha buain na monadh

In times gone by all over Scotland the early summer peat-cutting would be a good time (weather permitting). This from the Lowlands, a commentary on Allan Ramsay's *The Gentle Shepherd* of 1725:

> Meetings at the Moss, the provincial name of the place where peats are got, about Whitsunday [traditionally fifty days after Easter, so about mid-May], are the first busy, and joyous assemblages of the tenants and cottages in this district, after the hurry of seed-time is over; to dig, wheel, dry, and pile up for winter, the fuel it produces. It is carted home before hay-harvest begins. These mosses are, usually, so divided and scattered as to suit the general convenience, without disfiguring the country. Peat-spades, and peat-barrows, are necessary implements about every house; and a 'peat-stack' is, yet, a never-failing, annual, appendage to the farmsteads . . .

Now it is a job that fewer people do together. Cutters are from more immediate family groups. As people are educated first in towns, then cities, and work further from traditional crofts, and as sons and daughters marry partners from far beyond their villages and move away, the structure of life changes. George Mackay Brown wrote of a hundred or more men cutting along a single bank on the Orkney Islands. No more. In the course of one spring day on Lewis I met nine peat-cutters. Two men were on the moor in the middle of the island between Stornoway and Barvas – one had left the island as a young man and worked all his life in Glasgow, only now retiring to the family croft; the other, his brother, had stayed on the island. I met Barney and his family at Arnol, who manage the blackhouse museum, cutting peat both professionally to preserve the tradition but also domestically for their own use. But even among the keepers of tradition, the same issues arise: their son will be leaving for university in Glasgow in the autumn. I met a lone man cutting out on the Arnish Moor, and also working alone was a middle-aged man digging with a spade at Skigersta, at the end of the Western Isles Walk. His aunt and uncle, my friends Mary and John, both in their seventies, were cutting nearby.

Mary and John are fortunate. They live in the last house before the moor and can see their peat banks from the kitchen door; they go to the moor for about two hours every day for four or five days early each May to cut. That usually does them for the year – they have plenty still in the shed, which will more than last them to July/August, when today's cut peats will be dry enough to come in.

We drive down the gully, over the burn, then up the hill to the post marking their bank. We park up and walk across to it. Mary cuts, John throws. He wears big rubber gloves, Caterpillar boots. Mary and I are in wellies.

John starts at the right-hand side of the peat bank. He demonstrates 'turfing the bank', using a normal garden spade to take off the living top layer of vegetation. Two spades' width in from the bank, one spade's width along. The sod is deep enough to preserve all the roots intact and is thrown down onto the foot of the bank to allow it to re-root. This retains the living moor surface for future generations and for their sheep to graze. In this top layer, peat has not formed sufficiently yet and it is too

the Ness Historical Society archive. Her beautiful soft voice patiently reads the Gaelic names for the sections of the peat bank to me, her wee pupil sitting at the desk beside her. This treasure trove of memory is part community centre, part archive, museum, café, shop, tourist information, genealogist's gold mine and much, much more. In it are recorded the lives lived by this community down through the ages: from objects from peat carts, an early Christian carved stone and local lemonade bottles to cartoons of illicit underground whisky dens or bothans, sepia photographs of local girls gutting barrel-loads of herring, and local tales and songs by voices long fallen silent. It is not a sad, quiet place, however, but a vibrant, living, exciting hub within the community, and is currently being renovated with an ultra-modern design in a forward-looking project to secure its place in the community into the twenty-first century.

In another of the island's modern buildings – the school at Leurbost – preparations are underway for the new academic year.

Along from it there has been some serious peat-cutting, probably by at least four people, maybe more. I count five deep peats drying on the heather. Further stages of drying are evident with *rùdhan*, some combined into larger stacks and almost dry peats on palettes. Already some of the peat has been bagged, ready to be taken home. They are mainly in recycled builder's merchant and coal bags. Turning right on the road towards Callanish the hills of Harris loom from across the moor and lochs of this part of the island. There are lots of little off-roads to the peats, but very few signs of any being recently cut. This is a vast, near-empty landscape.

On the hillside of the glen there are finally some cut banks, so I stop to investigate. In this bog what is immediately noticeable is the wood in the peat. It doesn't appear to be too deep down but the first large piece is securely fixed, so is probably part of a large root. Feathery, stringy pieces of wood are poking through the peat scuffed into threads by the hefty boots of the cutters. The root and stump of a thin tree is near to a previous year's cutting. Further back another thicker stump hints at an uncovering a few years ago. The biggest root of all is being gradually absorbed back into the moor by heather, lichen, sphagnum. Sundews are sporadic, greedy to feast on the summer's latest hatchings before the insect-less winter sets in. I try to tickle one into closing with a stick of bog-cotton to no effect – no flies on this insectivore.

As summer draws to a close and island children grow into adolescence, play merges into necessary hard labour. I remember being struck by this as a child on a Highland summer holiday reading of Alasdair in Allan Campbell McLean's *Hill of the Red Fox*:

> We worked at the long, straight peat cutting I had first seen on my way to Achmore. The dry peats were stacked in small heaps, and Mairi and I filled sacks from these heaps. Murdo Beaton filled the big creel, working alongside us . . .

Transporting the Peats

In all peat-burning communities, school-aged children (seven to fourteen years) were employed in the task of bringing the peats home. Nowadays most banks can be reached easily by tractor and trailer. At peat 'home time' in late July or early August, the North Lewis moor is appreciably darker than the ground near the crofts. This is due to the combination of heather, which is in purple bloom at the time, and the blackness of the peaty earth where it has been exposed. The moor recedes in curving hills that give an impression of width and height – not a height of grandeur but of an undulating wave going on for miles and miles. Above it the sky is huge and as you progress further into the moor it becomes bigger. Climbing, the vista opens out and the third element that dominates this place comes into view – the sea. Huge sky, wide moor and the Minch in the east, the Atlantic to the north and west. The sky takes up 60 per cent of your vision, the moor 19 per cent, the sea 19 per cent and buildings 2 per cent. These buildings are predominantly crofts, but there is also the transmission station, water tower, lighthouse and the Free Church, with its bell-less bell tower.

Sometimes in the past the issue with transportation was not the distance from peat bank to home but the nature of the terrain between the two. It was not uncommon for the peats to be carried by creel on the backs of the women and children to a point where larger transport, whether it be horse and cart or, latterly, lorry, could complete the journey home. In Gaelic this is called *ag aiseag* – 'ferrying', suggestive of the watery condition of the land. Some of the most iconic pictures of peat are the Victorian photographs of multi-tasking women, knitting whilst carrying a creel of peat on their back. Such is the strength of this image that the very outline of the country has been likened to a woman with a creel on her back.

The creel bearer would be wearing a *cota*, a loose, knitted skirt that was rolled up at the back to form a soft pad called a *dronnag* to help take the weight of the peats and to prevent the woven basketwork digging into or rubbing the base of her back. An *iris*, the woven hair or bent grass supporting breast band could be ornately pleated or decorated, and as such was a status symbol even though it bound the carrier to her heavy load. These woven ropes were vital to distributing the weight of the peats without damaging the bearer's back, the harsh reality being that anyone with a musculoskeletal injury was a drag and a hindrance to the success of an intimate family group. In his *Lewis: A History of the Island* Donald MacDonald suggests that necessity of existence on a croft called for all the family, irrespective of age or gender, to be able to contribute to the physical work – to carry a load, to dig, to mind beasts, basically to help in all aspects of the transhumance and self-sufficient smallholding: 'Strong, sturdy, broad-backed women who could also help to push the boats up the beaches were appreciated.'

Out in the fresh mountain air of Harris there was still risk of injury. Coming down from high moorland Finlay J. MacDonald recounts that his father instructed him on how to carry a sack of peat, stressing the importance of letting it go should he trip – better spilled peats than a broken back.

In all peat-burning communities the children were employed in the task of bringing them home. In Shetland children were given responsibility for 'caa-ing' the horses. The stacking of the peat slabs in the creel was another learned art. Some were fantastically highly stacked in comparison to the depth of the basket, with a mere ribbon of string or rope helping secure them.

Most crofts would grow a willow – *caol* or *seileach* in Gaelic – protected from grazing animals by a stone enclosure and coppiced to provide canes for creel making; a mature willow may provide enough material in a year to weave two creels.

Another way to transport peats over rough ground was to use a wooden peat carrier, a device similar to a stretcher – two carrying poles, with wooden slats between, one person at the front, one at the rear. Nowadays you can still see wheelbarrows out on the moor and in the past a sturdy wooden-sided peat barrow would be used.

Horses were another way to get bulk peat home across boggy moor. A whole range of baskets, bags and nets woven from straw, bent grass, dried rushes or willow were used, manufactured at home during the winter months.

Of course some years when it had rained so much and the moor was saturated it was too wet even for horses to navigate, heavily laden with peat. The song 'Oran na Mona' from Islay tells of how the people had to carry the peat in sacks on their backs because the horses were just sinking in the mud.

By the mid-twentieth century Rodderick Campbell of Bragar on Lewis described as 'quaint' seeing in his youth an old woman leading a horse with two creels full of peat on either side. None of the women at the Thursday lunch club in Ness recalled seeing a Lewis pony, though one said that their Clydesdale 'Ginger' was the last heavy workhorse in the area. Its cart is now in the Ness Historical Society.

As well as to the home, excess peats had to be transported to where they could be sold.

In J.M.W. Turner's watercolour sketch *Peat Bog, Scotland c.1808*, drawn in the West Highlands, a pony is harnessed to a sledge. *Dreallag* is the

Gaelic name (originally meaning cat's cradle), with the peats slung in a rope-work frame between two poles. Sledges were also used on Shetland, and on the east coast in Angus as recently as 1963, where Henry Smith of Brigton was photographed with his horse-drawn sledge, mounted on which are two palette-sized squarish boxes with low slatted sides called 'slypes'. The hardy-looking ponies were harnessed one in front of the other, the first on long reins and the second behind, led by the bridle. The sledge has wooden runners. Even into the twentieth century remote communities like those on Fair Isle were still using ox-drawn carts.

Water could be a help or a hindrance. Sometimes the same burn would need to be crossed and re-crossed many times before firm land or road was reached. Salt water, though, could also be utilised.

In Shetland and other coastal communities boats could be used for 'flitting', as a way of either circumventing difficult country or because all the peat had been dug from their locale and they had to source it on uninhabited islands, as the people of Berneray off Harris did on Obisay, Votersay, Sartay, Neartay and Stromay. Christina Paterson would accompany Domhall Chaluim in his boat to collect peats at the end of the summer from these islands and, she said, 'I was as good at raising the sail as any man!' A special barge-type of boat was built by her grandfather, Ruairidh mac Ailein, called a *pearrdse*.

The people on the island of Tiree, when their own stock of peat was used up, had permission to cut it on the neighbouring island of Coll, or 20 miles away on the Ross of Mull. To transport the peat from the moor to the shore, they would either borrow local ponies or bring their own across the sea with them. On the Monach Islands the crofters would cross to North Uist to cut their peat, using shingle from their beaches as ballast to construct a jetty at Horisary. On the Shetland island of Fetlar it was a laborious process getting the peats home by land from where they were dug out on a peninsula, so they used the sea. Large families and a thriving fishing industry in the late nineteenth century meant that it was cost effective to bring the peats home by boat rather than across the moor. The process involved the building of various intermediate stacks – at the top of the sea cliffs, on the boat by the skipper who made sure the load was balanced, and on the beach when they got home, ready for ponies to cart the peat for final stacking for winter.

Even today, with mechanisation, the very condition of peat formation necessitates great care always be taken. Take too heavy a vehicle out onto the moor and it could either sink or seriously disturb the intricate balance between flow and retention of water that could destroy the fuel source that so many depend on. Peat roads are a vital part of maintaining that equilibrium, whilst allowing access to traditional banks far out in the moors. They are maintained by users, landowners, government, councils, community buy-outs and grazing committees, and, more frequently, wind turbine and telecommunications companies. Remedial works are needed every ten years or so, sometimes just a load of aggregate to patch up the places where water has washed the last lot away.

All across the islands, typical of the start of many peat roads are the metre or two of tarmacked rectangles adjoining the main road. These appear like spaced-out crenellations along the straight moor roads. The council road-builders take care of these, but once the tracks start over the moor it is the responsibility of the users. Occasionally other powers will step in to assist. In the 1980s when work was extremely hard to come by, the Manpower Services Scheme – which gave work in exchange for increased state benefits – funded the improvement and construction of many peat roads. This echoed previous schemes such as the Bridge to Nowhere and the construction of the Pentland Road, running across Lewis, linking Carloway with Stornoway in the east, though this was primarily to allow for the fast transportation to market of fish landed on the west. Perhaps the most famous of these work schemes in Scotland is McCaig's Folly, which still dominates the Oban skyline.

A relief fund was established in North Uist in 1923, as a very bad summer meant the crops didn't ripen; the potatoes were affected by blight and it had been so wet that the peats could not be carried home from the moor. To keep the people in work and give them an income, they were paid to construct peat roads. At Geshader in the west of Lewis the workers were paid in potatoes and meal by the Inspector of the Poor, and the road-building allowed the people to keep their pride. The road out to the peat banks there is known as Rathad a' Bhuntata – the Potato Road.

★   ★   ★

At Fivepenny Borve a patchwork brown, tawny and black peat stack is the last man-made structure before the moor. Peering into the cattle grid, I see my reflection in the water trapped at its bottom, imprisoned by the metal bars, and think – momentarily – of bog bodies. A shiver runs through me.

To leave the firm, solid land of 'normal' and venture out onto the moor is to cross a threshold and pass into a liminal world. The road starts out tarmac but soon becomes a track. Some friendly and inquisitive Belted Galloways take the time to examine me while a Highland cow unconcernedly grazes close to the crevasses opened up by the 'sausage machine' cutting that has recently taken place – the big modern tractor necessary to carry the chainsaw-like cutting blade can easily navigate this part of the moor close to the township, if it has double tyres fitted.

As the road becomes track, other tracks start to lead off from it and soon a network of paths traverse the bog, each leading to five, three, two and eventually to one family's peat bank. White-painted posts guide people on the safest route and away from someone else's bank. If you don't respect the moor, your actions can cause water to flow into a neighbour's peat. A recent problem is off-road vehicle joy-riding, which churns up the ground; inanely circling quad bikes and 4x4s create new pools and can destroy access for others dependent on the moor for fuel. Occasionally old tyres are placed to allow access over the remains of old peat cuttings. Burns have to be crossed either by fording them or building bridges. Some bridges are solidly built with stone and mortar, but so dank and lush is the moss that they are soon so colonised by sphagnum, heather and bog grass as to be indistinguishable from the surrounding moor. Others are of a more transitory nature, sheets of rusting metal laid over breeze-block piers. Sometimes after heavy rain the road itself becomes a burn, the grey stones and grit turned peaty brown by the overspill.

Building an embankment can help preserve the road; some sections are built up in a causeway 10 to 12 feet high, with steep ditches to control the water. This means that quite substantial bridges have to be built to support tractors. Such is the verdurous abundance in this moist, warm climate that most stone bridges become subsumed with unravelling ferns, sweet-smelling bracken and spongy mosses very quickly. Some bridges further down the slope are marked by wooden peg markers, so completely have they integrated with the vegetation. Generally at Ness the banks are well drained, sometimes with old peat cuttings used to divert water downhill. Over the years the change in water flow can force a change in the route of the peat road. Behind Habost two lines of bleached white stones stand stark against the brown/black peat as the route passes parallel to two disused peat banks; potholes are easily filled with living turfs. More old tyres see out their remaining years pressed into a particularly boggy stretch but often the tracks of those still attached to vehicles leave their prints across a quaggy stretch. It is not only tractors that leave their mark on the peat: sheep, cattle, deer, rabbits, dogs and their humans – locals and visitors – and, in a puddle, the webbed prints of white-fronted geese, those other returning North Atlantic tourists.

Nowadays peat is transported back to the croft in old tractors. If you are a Massey Ferguson tractor-spotter, then the crofts of Lewis are the place for you: 40s and FE35s from the 1950s, TO35s, 1080s, 135 Industrials from the '60s – all can be found. Nostalgia? Possibly. But apart from obvious reasons of thrift and prudent maintenance on an island 50 miles out in the North Atlantic, there are two other very sound purposes in keeping these half-centurians going – size and weight. As fields on the mainland expand to Ukrainian steppe or Canadian prairie dimensions, and demand ever-larger tractors, the geography and crofting system of the islands allows no such expansion; small is really all you need on small-holdings a few metres wide and a few hundred long. And if the other main activity is driving out over boggy moor, then lightweight is what you want.

Decline in peat-cutting creates problems for transportation. Now that fewer people cut, drainage becomes a problem, especially if your bank is located along a series of tracks that rely on others maintaining them. As Angus in Lionel explains, 'You relied on the whole community cutting their bank to allow water to drain from one to another all the way down to the sea.' Just as excessive action of joy-riding upsets the balance of water flow on the moor, so now does inactivity.

One of the striking things about both peat-cutting and transporting the peats home is the tight-knit family aspect; brothers working together, families cutting together, wives and husbands. Though personally I have not come across sisters cutting together I'm sure it happens. Mechanisation by tractor and trailer allows two people to provide for two households' supply of peat. It no longer needs sixteen or so trips to the moor and back to collect enough peats. More compact family groups and the use of electricity, gas, oil, etc. minimise the volume. The trips to the moor traditionally fitted in with school timetables, even for families with grand-parents living a traditional life on the croft and grandchildren on holiday from city schools, but also with modern working patterns. At Melbost Borve I got chatting to Donnie beside his peat stack. His brother works on the rigs off Aberdeen, so his long weeks on/off rota is flexible enough that he can help with bringing in the peats without being restricted to a set few days, as a 9-to-5 worker would be.

Once the peat reaches the main road, now the tractors and trailers carry on to the croft. In earlier times peat would have been emptied from creels and reloaded into carts. By the mid-twentieth century, lorries had replaced carts. A great piece of black and white footage filmed by Frank M. Marshall at Ness in 1948, now in the National Library of Scotland, shows men and women loading a lorry high with cut peat. The peats are thrown up from a pile on the roadside, deposited after coming off the moor by two men and four women. On the back of the lorry two men are stacking the thrown peats into a huge pyramid about 8 or 9 feet high. Just as there is an art to filling a creel to the max, so too a cart, trailer or lorry. This form of stacking peat has different names in Gaelic, as Anne Campbell describes them in *Rathad an Isein: The Bird's Road*:

> *steidheadh* – used to describe the construction of a stack of peat in such a way as to support the sides and shed rain
> *ladach moine* – a full trailer-load of peats, the sides of which are built up with *steidheadh* (like the outer walls of a peat stack)
> *a' tarraing na monach* – taking the peats home. Large loads can be constructed by careful placing of outer peats
> *taosgan* – a small trailer-load of peats which does not require *steidheadh*

Mary remembers her father building a wall of peats at the back of the trailer first, to stop the subsequent peats falling out, then building at both sides as the rest of the family tossed up the peat slabs to him. If someone was bad at throwing peats, the back of her father's hands would be bruised by being hit with the dried slabs. Most of this is unnecessary now, as peats are more often transported home directly from the bank in trailers with high sides behind tractors, with only the wall along the tail-gate end being built.

In Skigersta John has a small hopper attached to the back of his tractor rather than a trailer, but the principle is the same. Fifty years ago, ownership of a lorry meant plenty of hires at peat home time, and for shop owners like Donald Malcolm MacDonald of Kirkibost it was a valuable source of additional income. Nowadays it is not uncommon to see white Transit hire vans loading up with sacks of 'sausage' peat extracted from moors near the roadside in late summer.

Plastic sacks are now almost universally used in the transportation of peat, taking the place of the creel and more. Out at the banks, dried peats are to be found bagged up and awaiting uplift – bright yellow sacks with a green tartan pattern, a black-faced sheep logo, or the slogans 'Harbro', 'Clover Sheep Feed', or 'Macaskill Coal-Quality Fuel', with a scene of domestic bliss illustrated on the front, a roaring coal fireplace with a man reading the paper at one side and a woman on the other, a sleeping dog between in both white and blue sacks. These are recycled for peat, alongside builder's merchant bags and net rubble sacks – modern maishies. Some are lined up along the bottom of banks, others clumped together in pairs. Out at the peat banks on the moor between Stornoway and Barvas a grey domestic wheelie bin, laid on its side and turfed over, its lid secured by orange bailer twine, is the perfect repository for a stack of sacks waiting to be filled. Down at the roadside, sacks filled with peat await uplift beside an old battered traffic cone with 'Please park generously. Thank You' written in pen on it.

This newspaper article captures the scene of peats being loaded at the roadside in Skye eighty years ago:

Yesterday . . . I met an old man carting peat near Armadale. I stopped to examine the stack [built on the cart], which had all the appearance of a miniature Norman keep. A young native who was assisting to build the stack asked, on seeing my camera, if I would not take a picture of the old man and his horse.

'Why?' I asked.

summer's day; McLeods from Canada, Morrisons from New Zealand; coachloads of elderly passengers from the cruise ship docked for six hours at Stornoway harbour before sailing on to Iceland – Vikings in reverse; their younger, more adventurous compatriots in search of the 'real' Scotland on minibuses called 'MacBackpackers', 'Haggis Adventures', 'Rabbie's Trail Burners'; wind-beaten cyclists – their panniers dripping, legs sagging, hair sodden but spirits high as the final goal of their island-hopping quest nears its culmination; the migrant African swallows feasting on midges.

There are three peat stacks – two impressively large complete ones, and between them one under construction. When I was last there in June, just a small remnant of the previous year's peat stack was left; now these peats have all been burnt – the fire kept burning all year, as well as all day and night. An old red tractor with a trailer full of cut peats stands waiting to be unloaded to replenish the family's fuel stocks. The peats have been cut by Kenny and his brother on their bank on the moor behind the neighbouring township of Habost a few miles away; he reckons it will take about six trailer trips to build the stack. The trailer has been loaded in a very particular manner to make the construction of the peat stack as strong as possible. The hardest, blackest slabs of peat cut from the third or fourth layer at the bottom of the peat bank are put into the trailer first, then peats from other levels on top. When the trailer is tipped up, these peats are at the top of the pile; these he separates out. Being the hardest and densest, they make the best peats for building the outer wall of the stack, less given to shrinking or breaking in the face of the weather over the course of the seasons. The name for these outside peats, *sgiath*, Dwelly's Gaelic dictionary defines as 'shelter, protection . . . shield'. He puts these to one side in readiness to build the wall around the core of the jumbled other peats; the interior of the stack is not formally constructed.

He marks the line of the sides for each stack with pegs and blue nylon baler twine or fishing rope. The site itself is clearly defined by the brown peaty crumbles and dust from previous years' stacks and is on a slight natural slope towards a ditch; some heather transferred from the moor with the peats has started to grow at the fringes of this area, bringing moor to home. The stacks are situated here at the front of the house because he ploughs all the croft behind the house for potatoes; next to them is the telephone exchange building, which serves as a useful windbreak from the cold north wind. In the field beyond, two grassy mounds are almost exactly the same shape as the peat stacks, but concrete entrance doorposts and lintels mark them out as something quite different. Hollow where the stacks are solid, they are air-raid shelters constructed by the Air Force during the Second World War as part of the North Atlantic early warning radar system, protection against the German aircraft and U-boats that were the menace of so many island merchantmen and naval conscripts.

As his defence, Kenny builds his outer protecting walls in a herringbone pattern around the core of the stack – it is the strongest method of construction, as well as the most aesthetically pleasing. Starting at ground level, he builds using the narrow side of the peat slab with the clean, sharp-angled corners to the outside. The gap between the wall and the tipped jumble of peats is then infilled with softer, crumblier peats from higher up the peat bank so that the stack is as dense as possible and will not shrink or bulge (a tractor tyre is the favoured method of propping up a bulging stack, but Kenny's are so beautifully crafted that he does not need to do this). And so the stack is built, trailer-load by trailer-load. He will usually have eleven layers of herringbone wall and then 'thatch' over the top in a roof about six levels of peat slabs at its highest point. The finished stack will be about 18 feet long, 6 wide and about 5 to 6 feet high, with gently curved and rounded corners. Because of the quality of their fine construction and confidence of size there is almost a 'classical' quality to these stacks by which others can be measured. Their elevated position and proximity to the main road adds to their appeal.

Within the Ness area, there are other equally fine examples. Round the corner are Angus's peat stacks, almost a mirror image of Kenny's. I knock on the door. He's in the weaving shed at the back of the croft.

'Midges were too bad to continue building tonight, so I'm doing the tweed instead,' he tells me.

The weaving of the unique island fabric has been a retreat and comfort to many an islander, my father-in-law included, when the midges have been numerous and the other work too sparse.

There are three stacks next to the loom shed – one from last year, one from this and one under construction. They too are of a herringbone pattern. Eleven rows high, their structure curves round the corners, the tops made of stepped, flat layers. Angus also builds outer walls with the best hand-cut peats, herringbone for strength.

'Is that because you are a weaver?' I ask.

'No.' A joiner by trade, he has that very appealing Gaelic nature which modestly underplays his skill as a craftsman – he just tips other peats into the centre as he goes along, no special method, most stacks 'settle' fine. Doesn't cover or weigh down the stack. Will use a ladder to reach the top. What he does use are some machine 'peateater' cut peats. He says that they are slightly smaller than his peat-iron cut peats, which is why he doesn't use them on the outside, as they are not quite the right size. Even though it is a big structure, small differences like that matter. Three stacks of this size seems to be a common amount for seeing a family through the year. An old aerial photo of the township at the Arnol blackhouse shows houses with three big stacks alongside each other, probably about 5–6,000 slabs in each. Any leftover peats will be used to construct a small 'puppy stack'.

My father-in-law, also a joiner, told me about his experience of building. 'You would build your stack as close to the back door of the house as you could because in winter it was howling. I didn't have any special way of building the stack, only making sure it was sturdy and tight-packed to keep the peats as dry as possible. Some folk would build their stack in a special way – mine would be just normal but others would do fancy patterns. Some would be massive, and of course you'd get a bit of bosd [bragging] or even just a look. "My stack's bigger than yours", playground stuff. Sometimes you'd have "turfs" on the top rather than peats. What I call a "turf" is some of the ones from the layer near the surface of the moor. As well as keeping the peats underneath dry, you could use them to smoor the fire at night 'cause they were wet and grassy and slow to burn.

'In our day it was tarpaulins weighted with tyres to keep the stack dry, but before they'd use maybe a fishing net staked down or with stones tied on, like they had on the thatch roof of the old blackhouses, to stop the wind taking the stack away in a gale. In the middle of the stack you'd have a hollow for storing all the broken and crumbling bits of peat that couldn't be used in building the stack but would burn fine. Course you'd find all manner of things in the stack – creatures, our chickens would often lay in there and if you wanted to hide something it's a great place. Didn't they hide some of the bottles in the peat stack in *Whisky Galore!*? Compton MacKenzie down in Barra would no doubt have had a stack, though I don't know if he would have cut the peats himself.'

This story, from the archives at the School of Scottish Studies at Edinburgh University, was recorded in 1960 and tells of hiding spirits in a stack in South Uist. 'A boat from the Isle of Man came to Loch Aoineart, carrying rum. Local men took potatoes to the boat and brought back rum. The contributor's brothers decided to do the same, but hid their two bottles in a peat stack. The three of them were very keen to go to the peat stack when asked by their parents. However, when the rum was finished, their trips to get peat became a source of arguments, as they had been before.'

I came across one of those stacks that are weighted down with tarpaulins on the coast. One crofter told me that covering a stack with a tarpaulin would encourage condensation 'and you need the peats dry'; the wind and sun dries out a stack, even if the rain wets it. The top 'tiles'

his wooing, and unable to respect her decision he selfishly determined she should not be free to choose for herself. He waited at his window, armed with a shotgun, and when she went out to the peat stack by her house he fired, then turned the gun on himself. Fortunately, the girl survived.

Once built, the stack is almost immediately unbuilt: it is there to be burnt, but what of other forms of deconstructing? Peat is fuel and fuel is, in both senses, power, and as such, and given its location outside domestic confines, it is open to theft. In the language of the traveller people, the Beurla Reagaird, '*Biorach noib*' means '*Go and steal some peats*'.

In the Icelandic of another North Atlantic island with a peat-burning tradition the word for peat is *torf*. Nobel prize-winning author Halldór Laxness in his novel *The Fish Can Sing* describes a thief who steals a sack of peat from a man who gave him a sack of peat out of charity only the week before. Feeling remorse, the thief returns the sack he has stolen; however, the man tells him to keep it, but says, 'You have done something which God cannot forgive.'

Theft of peat was not restricted to Iceland. Back on the Scottish mainland, in the far north in Caithness 200 years ago, there lived a man known as 'Boustie'. He was a 'character', who was well known for stealing from his neighbour's peat stack. One night he was caught in the act of cramming peats into his sack by a group of high-spirited youths. Emptying out the stolen peats, they forced Boustie into the sack and tied the top. Now, at this time so-called 'Resurrectionists' were digging up the dead and in Edinburgh Burke and Hare had just been convicted of murdering people and selling their bodies to doctors for medical dissection, so this put an idea into their heads. The lads began to talk very loudly, so that Boustie could hear from inside the sack, that they were going to sell him to the local doctor for dissecting. Boustie was driven into fear and panic but, struggle as he might, he could not escape from the peat sack. Arriving at the doctor's house, they set the sack leaning against the door, rang the doorbell, then ran and hid behind some bushes in the garden. When the servant opened the door, the sack fell into the hallway of the house, with Boustie writhing and shouting. The commotion brought the doctor out from the dining room, where he was entertaining local dignitaries. At that moment, Boustie's struggles paid off and he burst from the sack, screaming obscenities and threats of setting the law on the murderous doctor before vanishing at great speed out into the night. It was said that after this incident Boustie was cured of his peat-stealing.

Each slab of peat that makes up a stack contains a memory or a story. In constructing a stack these individual slabs are vital and whilst the key exterior ones that we see looking from the outside may seem to constitute the structure around which the stack is built and hold it together, the inner jumble of peats – some also perfectly formed but many imperfect, broken, partial or crumbled fragment – is the body of the stack without which the whole thing would be nothing but an empty shell.

A scent of the peat reek can be detected at the fringes of the big, transformational events in the nation's history: a sixteenth-century post-Reformation inventory of Kilwinning Abbey lists '9 fathoms of peat stack' among the rentals; in early 1745, while the Jacobite Earl of Wigtown is selling off lands to raise funds for the upcoming Rising, his yearly rents include '60 Loads of Peat'.

A Presbyterian may also point out, with a dour Scots fatalism, that whatever the importance of peat slab or human, we all end up the same: dust and ashes.

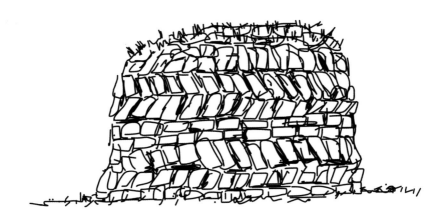

Littoral: Marriage

A summer island wedding is a great thing. The most famous of peatland weddings is that of Mairi:

Mairi's Wedding (The Lewis Bridal Song)

Chorus:
Step we gaily, on we go
Heel for heel and toe for toe,
Arm in arm and row on row
All for Mairi's wedding.

Over hillways up and down
Myrtle green and bracken brown,
Past the shielings, through the town
All for the sake of Mairi.

Red her cheeks as rowans are,
Bright her eye as any star,
Fairest o' them a' by far
Is our darling Mairi.

Opposite: *Peat Creel.*

Plenty herring, plenty meal
Plenty peat to fill her creel,
Plenty bonny bairns as weel
That's the toast for Mairi.

Bog myrtle, shielings, plenty peat. A Lewis idyll – only it was composed in twentieth-century Glasgow by Sir Hugh Roberton, conductor of the Orpheus Choir.

Only later came the realisation that it was as authentic as Brigadoon, the mythical Scottish village that appears out of the Highland mists. The original song was written in Gaelic by South Uistman John Roderick Bannerman (1865–1938) for Mary MacNiven on the occasion of her winning the gold medal at the National Mod, the annual Gaelic singing competition.

My friend Fiona told me of a hen party she attended in Shetland where they visited all the bride-to-be's childhood haunts: cakes in the bakery she used to go to on the way to school, pseudo peat-cutting at the family peat bank.

A Highland wedding tradition was the washing of the feet. Hector Sutherland from Brora (1871–1963) tells of the night before the wedding:

There was 'the washing of the feet', they called it. The coming bridegroom had to put his feet in a big tub and they were washing and there were silver and rings put in it, and diving, and who would get it, and the person to get the ring was the first to get married. Then they had a piece of peat and, of course, they broke the bits of peat and they put it in the tub, then they were pickin' a bit each an' oozin' a bit out to see the colour of the fibre and whatever the colour of the fibre in your bit [laughs] that's the colour [more laughter] of the hair of the one that you are going to marry . . .

So in this role at a wedding, where bride and groom are looking forward to a long life together, peat becomes not a preserver of things from the past but a foreteller of the future. A secure future is a consideration in making any marriage decision. Jockey in this traditional Scots rhyme not only has gold and gear to offer Jeany but a stack before his door to 'make a rantin fire', metaphorically or no.

Jockey Said to Jeany

Jockey said to Jeany, Jeany wilt thou do't?
Ne'er a fit, quo' Jeany, for my tocher-good,★
For my tocher-good, I winna marry thee.
Ev'n ye like, quo' Jockey, ye may let it be.

I hae gowd and gear, I hae land eneugh,
I hae seven good owsen ganging in a pleugh,
Ganging in a pleugh, and linkin' o'er the lee,
And gin ye winna take me, I can let ye be.

I hae a good ha' house, a barn and a byre,
A stack afore the door, I'll make a rantin fire,
I'll make a rantin fire, and merry we shall be;
And gin ye winna take me, I can let ye be.

Jeany said to Jockey, gin ye winna tell,
Ye shall be the lad, I'll be the lass mysell.
Ye're a bonny lad, and I'm a lassie free,
Ye're welcomer to take me than to let me be.

★tocher – dowry

Autumn

THE AGE OF BRONZE

Burning the Peat: The Hearth

Peat, having come from water to tiny bryophyte sphagnum on the moor, then been formed by humans into a solid structural stack, is about to enter its final metamorphosis and change into smoke. Between the dried peat stack outside and the hearth at the centre of the home lies the final journey.

Midway is the entrance, the threshold between out and in, between surviving the year in this landscape and not. In the porch area of a traditional Lewis blackhouse, space was set aside for storing peat to dry off beside barrels with the house's water supply. To enter one of these homes is like but unlike. The construction of the walls makes the entrance passage longer than you are used to, with both an outer wall and an inner wall of big, dry stones infilled with earth and turf for insulation. The door is low and you stoop to enter. The tendency is then to straighten up once you have taken a step or two, but, because of the depth of the walls, if you do so, you bang your head on the stone lintel.

Once inside, the darkness of this blackhouse adds to your initial concussion, for there are no windows, only, perhaps, a skylight in the thatch. And in that small opening (glass being a rare commodity) it was not unknown for a skate skin to be used as a pane. To complete your

Part title illustration: *Cast-iron Stove.*

sensory disorientation, the smell and smoky atmosphere of the peat fire burning in the central hearth brings water to your eyes.

What then of burning the peat? Whilst not providing the energy of other fuels, its main benefits, given the right conditions, are its relative accessibility and low cost in both labour and monetary terms. Archaeology tells us that man will always utilise what comes to hand, and while wood is undoubtedly the easiest fuel to collect and burn, it is a finite resource needing years to regenerate. Early agrarian societies had to move when the forest became exhausted and farmland unproductive. Whilst probably mined since Roman times, coal is difficult to obtain, sources near the surface are extremely rare and before the Industrial Revolution of the late eighteenth and early nineteenth centuries technologies did not exist to make it widely available. Oil was unattainable.

Peat, on the other hand, was readily available. How was it discovered that peat was flammable? Presumably a natural fire, caused, perhaps, by a lightning strike which literally set the heather ablaze, revealing that the peat underneath burnt well. Instances of lightning striking on peat moors are recorded in more recent times, sometimes fatally, as when two young women sheltering by a stack of cut peats were killed at Ecclefechan in Dumfries in the early nineteenth century.

An anecdotal story of how peat burning for fuel began in Scotland tells that when wood ran short on the Orkney Islands in about 900 CE the Viking overlord Earl Eyner told the populace to cut turf and dry it to use for fuel. This proved such a good substitute that the practice spread all over Scotland.

At the Scottish Crannog Centre on Loch Tay there is a great demonstration of how to light a fire using the Iron Age technology of a bow drill to generate heat. It rotates a wooden spindle to create enough friction to make an ember, which can then set alight some deadwood or sawdust to kindle a fire. Another traditional method, used by the Vikings, was a tinderbox, perhaps with a piece of horse hoof fungus, often found growing on dead or dying birch trees on the lagg fen. Four pieces of fungus were even discovered among the possessions of Otzi, the 5,000-year-old hunter found preserved in the ice of the Italian/Austrian Alps in 1991.

The best way to keep a fire is not to let your old fire go out. In the peat culture of Scotland this was invariably the case. In some parts it still is.

I walk past the crofts of Ness on an autumn afternoon and the smell of the peat fires is in the air. The mother of the house is out at the peat stack – anorak, headscarf, plastic bucket and hammer: the tools of the trade for replenishing her hearth. We chat about the stack and she says regretfully that her adult son is out on the moor bringing his beasts home for winter, otherwise he'd have been delighted to talk about how he built the stack. As I depart she gets back to the job in hand and, picking a big, black slab of peat from the stack, wallops it with the hammer and collects up the pieces in her pail, ready for the fire. Further along the road a heavy cast-iron knife-cum-bludgeon lies beside a stack, ready to break up the peats.

Constant burning keeps the fire bricks in the hearth radiating heat and if, as many do, it also heats your hot water boiler, there is a double benefit. In Lewis, where there are still many power outages, not having to rely on electricity is a great advantage. In the peat culture the fire would be deliberately kept on all night. This is known as 'smooring'. To smoor a fire, a slow-burning peat slab is covered with ashes to keep it smouldering throughout the night. My mother-in-law would wet the peat to keep the fire just in but not let it go out. Some people use the peats from the top of the stack for this, those which have absorbed the rain, or use the damper, grassier peat cut from the top layers of the peat bank, which burns less well – sometimes a combination of the two. In the Catholic households of the peatlands a prayer dedicated to the Virgin Mary or Saint Brigid was, and still is, said to keep the fire alight during the darkness, the ashes drawn in with four sweeps like a cross.

This prayer is from a Catholic household on the island of South Uist:

Smooring the Fire

I shall smoor the fire to-night,
As Mary would smoor the fire.
Who shall be on watch to-night?
Bright Mary and her son,
And a white angel at the door of the house
Until to-morrow comes.

In a Protestant Church of Scotland household the fire would be kept going all day on Sunday, but in a Free Presbyterian one the fire would be banked up on a Saturday night and not touched on the Sabbath, then rekindled again on the Monday morning. No cooking would be done either, so any food and drinks taken would be cold.

Here are the impressions of peat fires by the American Margaret Fay Shaw from the 1930s:

The two chimneys at either end with open fires burn peat whose pungent aroma permeates everything in the house ... The house had originally been a *taigh dubh* or blackhouse, which is the oldest type of dwelling in the Hebrides and now rarely seen. Then the inside walls were not lined and the place for the fire was laid on the floor. The opening in the thatch for the smoke to escape was never directly over the fire in case of rain, but fire and door and chimney-vent were made according to rules that would draw the smoke out. But wind and rain will defy many better chimneys, and at times the smoke would gather in the rafters, and to avoid it one had to sit on a low stool. Peat smoke is not harsh and is said to be disinfectant. The people have been accustomed to it since fires under cover began. The pleasant thing about such a fire was that there was room to gather round it in the true sense. The light was then the crusie lamp or *cruisgein*. The fuel was fish liver oil, and the wick was made of the pith of rushes or *luachair*, which was dried by the fire and then plaited.

You can gain just a little insight into the effects of a peat fire burning in a hearth in the centre of the floor by visiting the blackhouse at Arnol on Lewis. Here is a verse of a spinning song:

Cuigeal na maighdin,
Cuigael na maighdin,
Stocain air bhiorain,
An t-snighe 'ga froighneadh.
Cuigeal na maighdin,
Cuigael na maighdin.

Distaff of the maiden,
Distaff of the maiden,
A stocking on knitting pins,
while sooty drops ooze (from the rafters).
Distaff of the maiden,
Distaff of the maiden.

The sooty drops – *snighe* – form among the rafters of the blackhouse because of the variable ventilation and, when the chimney wouldn't draw, the smoke would gather in the roof space. This, as well as the low doorway, could pose problems for tall visitors. After landing on Eriskay off South Uist on 23 July 1745 Bonnie Prince Charlie was entertained in the house of a tacksman, Angus MacDonald, where a peat fire was burning. Being royally tall and the house roof pleasantly low, the 'lad born to be king' found himself having to go outside for constant gasps of fresh Hebridean air.

In terms of the mechanics of cooking, the rope, or latterly an iron chain, above the open fire, which was adjustable by using an 'S' hook, was moved up and down to decrease or increase the heat as required.

In search of a serving girl at the Inveroran Inn near Bridge of Orchy in 1803, Dorothy Wordsworth opened the door of the kitchen. She recorded in her journal:

> About seven or eight travellers, probably drovers with as many dogs, (they) were sitting in a complete circle round a large peat-fire in the middle of the floor, each with a mess of porridge, in a wooden vessel, upon his knee: a pot suspended from one of the black beams was boiling on the fire; two or three women pursuing their household business on the outside of the circle, children playing on the floor. There was nothing uncomfortable in this happy confusion: happy, busy or vacant faces, all looked pleasant; and even the smoky air, being a sort of natural indoor atmosphere of Scotland, served only to give a softening, I may say harmony, to the whole.

As the nineteenth century brought Lowland wages, trade and commodities to the peatlands – as it was doing across swathes of previously isolated, semi-independent or autonomous communities right across the British Empire – the central hearth blackhouses began to give way to the more cottage-like 'white houses', with windows and hearth (or hearths) moved to the gable ends, with their chimneys and fireplaces integrated into the structure of the walls. Sometimes the chimney stacks were of wood, as in the reconstructed houses at the Highland Folk Museum at Newtonmore. Cast-iron fire equipment from the industrial Central Belt became common and some fireplaces would be equipped with a 'swee', a crane-like device for swinging pots in and out of the fire. Heavy-duty pots and big black kettles were imported from the Lowlands, along with the big black metal stoves which replaced many an open hearth. Fine examples of all three can be seen in the croft opposite the blackhouse at Arnol.

A striking feature of a rural island kitchen was the kettle, always on the boil; tea consumption was a staple of life, especially strong, well-boiled tea, as brown/black as peaty water. As early as 1773, Dr Samuel Johnson and James Boswell were comforted by a cup on their tour on the island of Coll:

> It was very heavy rain, and I was wet to the skin, Captain M'Lean had but a poor temporary house, or rather hut; however, it was a very good haven to us. There was a blazing peat-fire, and Mrs M'Lean, daughter of the minister of the parish, got us tea.

Tea is universally offered when visiting on the islands. My father-in-law still takes his tea very strong and if made with a tea bag he will leave it in the cup to make it peaty black. He acquired the taste from Lewis, where, when replacing windows or doing joinery in people's houses, he would be supplied gallons of the stuff. Hospitality is taken very seriously and he would also be offered food whilst modernising old houses, adding windows and skylights. He recalls among older people being offered, and of course accepting, dinner, this often being two bowls placed on the table, one of home-salted herring and one of boiled tatties. No cutlery. 'The best way to feel a fish bone is with your hands.'

Modern multi-fuel stoves are popular on the peatlands, giving flexibility and with no drafty chimney, though the old people loved their big black stoves and said they couldn't get the same control of the heat for baking scones or pancakes. I suspect the same conversations took place when the black stoves replaced the open hearth.

By coincidence, as I was visiting the blackhouse museum at Arnol, along came a lorry with palettes stacked with white, yellow and black plastic bags of coal, smokeless coal, red net bags of logs and kindling. It is these coal bags that, once emptied, are universally used to carry peats from the moor to the stack.

Then, with the peat reek from the constantly burning fire inside the museum still in my nostrils, along the road I come across a reconstructed Norse water-powered mill, like a mini blackhouse, towered over by a newly built wind turbine.

Another day, over on the east coast cliffs at Tolsta, I was taking a photo of a peat stack with the Minch in the background when a Scottish Fuels oil tanker drove past. I followed it down the track to the house to which it was delivering. The driver had already started to pump the oil through a large red hose into the green plastic storage tank so ubiquitous in most of the front gardens of modern crofts. The noise of the pump was too loud for conversation, but once he'd finished we chatted about the oil, coal and peat. Business was good, most people now used oil as their main source of fuel, but many used a mixture too. As if to emphasise the point, parked round the corner was a white van with ladders on top and on the side: 'Zoot and Zweep, Chimney Sweeping Services'. Chimneys where peat and coal are burned quickly build up bituminous resin and need swept twice a year – given that most buildings on the islands are only one-storey high, many people sweep their own with rods and brushes.

The build-up of soot in the old houses of the North Atlantic was universal. Icelandic author Halldór Laxness describes an old lady thus:

I cannot remember her otherwise than bowed and toothless, with a bit of a cough and red-rimmed eyes from having to stand before the open fire in the kitchen-smoke of Brekkukot, and before that in other cottages whose names I did not know. There might sometimes have been a little soot in the wrinkles of her face ...

Brekkukot was based on a real farmhouse, Melkot, built in the eighteenth century. It was a double-gabled traditional Icelandic turf structure. The links between these North Atlantic treeless island communities are strong. Norse still permeates the place names of Lewis, whilst recent genetic studies in Iceland have revealed an ancestry of male Viking DNA mixed with that found in the women of the western coastal fringes of Scotland.

The blackhouses that survive today, or that have fallen out of use only within the last two generations, are not that old but a variation on an ancient theme. Beloved by the current age for their 100 per cent use of local materials and insulation, most only date from the late nineteenth century when, following the Crofters Act of 1886, people had enough security of tenure to actually invest in housing that wasn't going to be 'appropriated' by a scavenging landlord. Poor-quality housing marked the period between the end of the paternalistic clan system in the mid-eighteenth century and the Act, and although living conditions in the peatlands were better than those of many in the vastly expanding Lowland cities, they were not immune from the diseases of poverty and lack of proper sanitation. Living in smoke-filled accommodation shared with livestock and no running water saw high infant mortality, many mothers dying in childbirth, or fatalities from TB, whooping cough, scarlet fever and measles.

One man who both cared passionately about the poor and suffered from TB was George Orwell. In 1946 he moved to the Hebridean island of Jura, seeking peace after war to finish his masterpiece, *1984*. His biographer Gordon Bowker recalls in *Inside George Orwell* a letter he wrote of the atmosphere there: 'It is not a cold climate here ... actually the mean temperature is probably warmer than in England, but there is not much sun and a great deal of rain.'

Perfect weather for the growth of sphagnum mosses, not so good for someone dying of tuberculosis.

Whisky

Would Orwell, even in his final days, have imagined a dystopian future where the Jura distillery would produce in the year 1984 a whisky in his honour, limited to 1,984 bottles?

Edinburgh airport in September after the end of the Festival and in the duty-free shop is a wall of whisky, stacked with row upon row of bottles. Bargain deals on mass-market blends for £15, exclusive hand-crafted wooden boxes lined with tweed containing limited edition malts for £3,500. No matter the price, what is on offer here is Scotland, distilled. For the Scot leaving, a reminder of home; for the homeward-bound, a memory of a now fast-receding present.

Imagine, if you will, my wife and myself in Salisbury, southern England, opening a bottle of Bowmore whisky. Angie's work had taken her there, and though we still lived in Edinburgh she would spend weeks away at a time. The aroma and then taste of this seaside peaty malt immediately transported her to the Hebrides, from Stonehenge to Callanish. For her, the combination of the sea, the whisky and the smoky peat brought back memories of home in another land.

Said to be the oldest licensed distillery on the island of Islay, Bowmore was established in 1779 and, like all the island whiskies, is known for its strong peaty taste. In the past, most would gradually move on to drinking Islay malts after first trying blends, then softer, gentler Speyside malts –

this was certainly my route, almost literally. But today the peaty malts are the most popular – perhaps, experts suggest, because it offers an 'authentic' taste, which appeals to an adventure-seeking young whisky drinker who appreciates full-flavoured foods, whether extra virgin olive oil, homemade farmhouse cheeses or craft beers.

As a young teenager our annual Boy Scout summer camps seemed to regularly involve a trip to a distillery. At Aberlour on Speyside not only did we visit the distillery but also engaged in a uniquely Scottish form of Scout pioneering – rescuing whisky barrels washed into the torrent of the river when the Spey, in spate after torrential rain, washed them away – tripods lashed together, knots, pulleys and my first smell of the inside of a sherry cask. Another year we were on Islay, and Laphroaig was the distillery we were taken to. If I remember correctly, the Scout leaders very kindly offered to look after the free miniatures we received as part of our admission price on the guided tour. I can only put it down to this excellent example of combining Lord Baden-Powell's passion for outdoor adventure with appreciation of the fruits of generations of Scottish ingenuity that my brother has just completed a cycle tour of all the Islay distilleries forty years on. Gordon Caskie, from a famous Islay family and one of our far-sighted Scout leaders, now eighty-five, was born and raised on the moor at Glenegedale, near the airport, where the peat is cut for the distillery malting.

These Islay distilleries love the smoky peats cut from the top of the bank, going through about 2,000 tonnes per year. It is the burning of the peat to dry malted barley in a kiln that makes island whiskies distinctive. Given that virtually all the Islay distilleries use pure rainwater that has passed through the peaty filter of the sphagnum, it is doubly peated.

By chance, my two favourite malts mentioned here, Bowmore and Laphroaig, have their own small maltings, as well as using the barley malted commercially at Port Ellen, which supplies the island's distilleries. Laphroaig proudly displays 'since 1815' on its bottles and, as seen above, we now associate peated malt with the Highlands and Islands, but this has not always been the case. Just as in peat-cutting, the practice was once widespread across Scotland.

The aroma of burning peat has been described as having antiseptic qualities – like bandages being manufactured from the absorbent sphagnum moss. I can still clearly remember attending my first-ever whisky tasting and, naturally, being attracted to try a Laphroaig malt after my Scouting visit. TCP, hospital ward, disinfectant – sounds weird that these could be good flavours in a drink, but they are – peaty, smoky, delicious! And after my dram is drunk, nosing the glass: it smells like a miniature peat fire has been burning in there. The contrast between the delicate smell of the moor and the strong reek from the peats in the whisky is like the difference between ingredients when baking: the flour, eggs, sugar and butter are almost odourless, but the aroma of the baked cake is delicious.

Like baking, creating a whisky is chemistry, and that process has a language of its own, just like the names for the plants that constitute the peat or the different words for cutting it on the moor. It's not the purpose of this book to be a study of whisky-making, but here are a couple of terms which you may encounter and particularly apply to the peatiness in your dram:

Phenols – in essence a range of chemical compounds based around carbolic acid (C_6H_6O) but used in the context of whisky to describe the peaty flavours and aromas.

PPM – parts per million, used to measure the peatiness of a whisky, e.g. the specification for many Islay malts will be between 30 and 50 ppm phenols.

The making and selling of whisky is a modern, international, multi-billion-pound business that generates huge revenues for the companies producing it and for the governments that tax it. Its roots are as a 'cottage' industry produced domestically on a small scale, and whilst the remnants of the culture that produced it still survive, would it be naive to presume that it doesn't still exist in that form? The moors are big, dotted with semi-habited shielings and hidden places. Does that curl of smoke signify that Alasdair is drying his socks by the fire, or making a cup of strong, black tea . . . or brewing something stronger?

The novelist Neil Gunn was a lover of whisky. In *The Silver Bough* he describes an illicit, moorland drinking den inside a remote wheelhouse (an underground Iron Age storehouse) and in his classic 1935 *Whisky and Scotland* recounts visiting a friend from Caithness who produced a homemade bottle containing whisky that had passed down through his family and was 104 years old. Intrigued by the chance to savour a true Highland throwback made before industrialisation and marketing men dictated taste, they open the centurion bottle:

Let it be said at once that the liquor in that bottle was matured to an incredible smoothness. I have never tasted anything quite like it in that respect, yet it had an attractively objectionable flavour, somewhere between rum and tar, to our palates. I suggested that the malt must have been dried over a peat fire into which some ship timber – probably gathered on the Pentland shore – had been introduced. But my host hit on what undoubtedly had happened. In these old days, when it was the custom to have a fire in the middle of the floor with a hole in the roof for the chimney, some specks of the old glistening soot from the rafters had fallen amongst the malt. Perhaps during the first few years of maturation this might hardly have been detected – or, again, may have been deliberately aimed at as an elusive part of the whole flavour! – but certainly in this extreme age it was all-pervasive; it had become, indeed, the spirit's very breath.

Peat has that retentive quality. In the digging of it out of the moor you see its timeline stretching back over thousands of years, from fresh buds sprouting on the living turf through the older vegetation down into the roots; the gradation of tone and colour from ochres to umber to chocolate and rich black-brown marking the passing of centuries and millennia of sphagnum – tiny individual lives making up a huge community compressed by time.

In burning this and releasing the peat's aroma there is an immediate sensory connection with the past. Memories of people and times, of generations now gone, are conjured up. For those still living on the peatlands, there is a two-fold element of honouring those who have come before.

First, there is a physical connection in cutting on the family peat banks, which not only your parents and grandparents cut but which were shared in an annual ritual by your wider community. Whether your family had had that croft since a land-raid post-First World War or since your people had been cleared off ancestral lands to make way for sheep in the nineteenth century, or it had been in the family since way back, there is a ritual there in performing the same actions with the same tools – cutting, catching, throwing – that is almost meditative on this open expanse of moor under huge skies, under the eyes of your god.

Second, there is the burning. By doing this you are not only releasing the smell of your childhood, of people's houses that you knew, of your own memories, but also the memories of everyone who grew up in your community and those who have always lived in your community. The slightly disinfectant tang holds an almost curative quality; there is a sense of cleansing and renewal.

As the smoke escapes through the chimney, the private becomes public and, walking past, you smell the reassuring and wonderful mix of peat smoke and salty Atlantic air that is as rich as a glass of Bowmore Islay whisky. No matter what time of year, many home fires are kept burning, just as in the past. You also connect in a domestic way to the wider *Gàilteachd* and similar communities who share your culture and belief systems.

> *Tha a smuid fhein an ceann gach foid,*
> *'S a dhorainn ceangailte rig ach neach.*

> Every peat has its own smoke,
> And every person has his own sorrow.

In days where the fog is down and the ghostly sweeping beam of the Butt of Lewis lighthouse highlights the nothingness and its foghorn boom reverberates like a footstep on the quagmire moor, the smoke hangs heavy, close. But on a clear day with a brisk wind you fancy that the peat smoke can reach all the way to America, twitching the genetic Scots noses of emigrant communities in Nova Scotia, South Carolina and into the West and beyond.

1. *Above.* The bog is half-land, half-water. Sphagnum mosses hold twenty times their own weight in water. Stop moving and you start to sink!

2. *Right.* Another length is added to the peat corer on Kirkconnel Flow.

3. Paradise Garden: Scarlet-cup lichen, the flaming red-tipped *Cladonia coccifera*, grows abundantly and joyously on the exposed black peat; fluffy white buds of bog cotton, *Eriophorum angustifolium*, spot and wave in the wind like a cheerleader's pom-pom; Moorland spotted orchids, *Dactylorhiza maculata*, growing on peat at 2,000 feet on Cairngorm mountain; jewel-like round-leafed sundews, *Drosera rotundifolia*, are beautiful when discovered but not so for all – it is perhaps the most well-known insect eater on the bog!

4. *Right.* After a long, claustrophobic North Atlantic winter spent indoors, spring and summer at the shieling on the moor was a time of happiness and joy. (Steven Sheppardson/Alamy Stock Photo)

5. *Below.* Stag Bakery van: More than just a mobile shop, the baker's van brings people in isolated communities together for exchanges of news and companionship. Note the peat stack behind.

6. *Left. Tairsgeir* and spade: The tools of the trade. The long-bladed *tairsgeir* (or peat spade) and a spade for 'turfing' out on the moor.

7. *Below.* Preparing lunch in a tent: Post-Second World War army surplus tents ensured that even in the worst spring weather cutters could enjoy a dry lunch out at the peat bank. (Comunn Eachdraidh Nis)

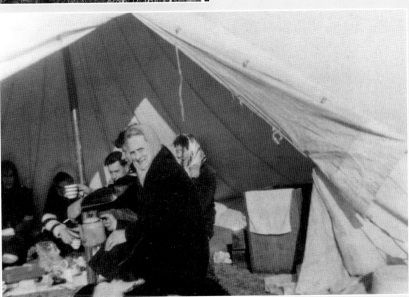

8. Mary and John at their family's peat bank. Cutting, catching, throwing; cutting, catching, throwing . . . the rhythm of peat-cutting weaves itself back through the ages.

9. From high above, the Lewis moor looks still, but under the peat the bituminous oils are moving. Having retreated before the winter's chill, they have joined that secret underground blossoming, invisible beneath the unchanging sphagnum, the season-less lichen, that heralds spring. (© Patricia & Angus Macdonald/Aerographica)

10. Peats drying: *Rùdhan* (three slabs together with one on top) on the east Lewis moor. Across the Minch lie the Assynt mountains of the mainland.

11. Lines of drying peat on the moor: The gap between the top of the bank and the cut peats is called *rathad an isein*, 'the bird's road'.

12. A cairn on the moor behind Cross, Lewis: The grey-white stones are heavily stratified with black, layer upon layer laid down over geological ages, making the soft sponge of the peat blush with its youthfulness.

13. At peat home-time all join in: This wooden tractor and peat trailer is in a garden at Eoropie at the very north-western tip of Lewis. Behind it the family's peat stack, then the North Atlantic Ocean.

14. Images of multi-tasking women, knitting whilst carrying creels of peats on their back, are some of the most iconic of late nineteenth-century Scotland. (Comunn Eachdraidh Nis)

15. *Peat Bog, Scotland*, J.M.W. Turner, *c.* 1808: Despite the romantic setting, Turner depicts the sheer hard labour and toil in the lives of the peatland subsistence crofters. (© Tate, London, 2018)

16. *Right.* Traditional and modern fuel sources on the moor: Sacks of hand-cut peat await transportation home under a wind turbine. (Ashley Cooper/Alamy Stock Photo)

17. *Below.* Kenny building the third of his peat stacks, Lionel, Lewis. The herring-bone method of construction echoes the tweed woven by many peat cutters.

18. *Left*. On entering a blackhouse, you are enveloped by the 'peat reek' from the hearth in the middle of the floor.

19. *Below*. Bowmore Distillery, Islay: A piper would lead the peat cutters out to the moss in spring. They would cut one load of peat for the distillery, one for the landowner, and sell a third to people of the town for a shilling per cartload. (Shutterstock)

20. John Kay's caricature of Lord Kames (left). Kames offered large tracts of Blairdrummond Moss for 'improvement', reclaiming the peat moors for agriculture.

21. *Women working on the Peat Moor*, Vincent Van Gogh (Nieuw–Amsterdam, October 1883): The back-breaking labour of cutting peat in the Dutch landscape. (Oil on canvas, 27.8 x 36.5 cm, Van Gogh Museum, Amsterdam (Vincent van Gogh Foundation), s129, F19.)

22. *Above*. Digging trenches, digging graves: Work with spades continued for the peatlanders in the hell of Flanders fields during the First World War. (Colin Waters/Alamy Stock Photo)

23. *Left*. Etched in the window of Croick Church by crofters betrayed and cleared off their ancestral lands: 'Glencalvie people the wicked generation'. (Historic Environment Scotland)

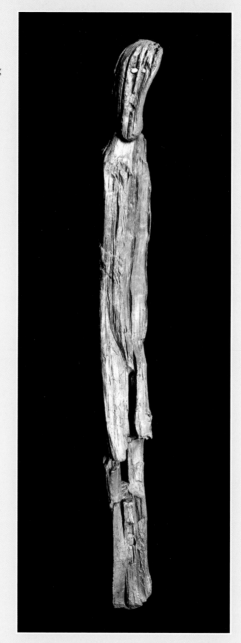

24. The Ballachullish goddess, unearthed by a peat-cutter. She stood in woven bower at the ferry crossing, accepting offerings from peatland travellers. (© National Museums Scotland)

25. Passing through the gate at the end of the year, looking back on Rannoch Moor.

Highland Power

In my thirties I started hillwalking. On the first Sunday of every month, together with my brother and a group of our friends, I would set off early and head north, taking up the challenge of climbing all 282 of Sir Hugh Munro's mountains over 3,000 feet. Each of us had different reasons for going: some wanted to chalk off another ascent or get to the top as fast as possible; for others, keeping up with the latest equipment and technology was important. Whatever the reason, for us all it was an exhilarating change from daily work or home routine. I think we all enjoyed the social side – a 'rusty nail' (malt whisky and whisky liqueur) from the flask at the summit, the pub at the end.

I think my interest peaked one June weekend when we climbed the south Glen Shiel ridge – seven Munros in a row above the road to Skye. It was arduous but not exceptionally so; the weather was hot and the views vast and magnificent, the sense of satisfaction high and the drinking afterwards monumental. Perhaps I felt that after such an achievement it would never be so good again and best to stop. There was also a balance to be met between the needs of us as a group and as individuals. I had always been more interested in the journey than reaching the top; for me, it was the nature, the history, why the mountains and the lands are as they are. As we climbed I never felt I had enough time to stop, look and listen.

On a dreich autumn morning I am back on that road from the Isle of Skye to the Great Glen just south of the ridge. Today all is wet and tawny – four times more rain falls here than on Scotland's east coast. Squelching under my boot, the moss is saturated and lush beneath grey late September skies. Across an auburn bowl of bog cupped within the hills rushing towards me is a mountain stream; it powers past me in torrents, foaming into white waterfalls and cascades down towards Loch Garry 1,500 feet below. For this is a peatland too, bogs forming in pockets at any altitude across this landscape. All around are huge vistas over mountain, moor and loch: to the south the Loch Ness monster, ahead the Jacobite battlefield at Glen Shiel, Eilean Donan Castle, the Isle of Skye, the fairy flag at Dunvegan – the epitome of Scotland to many around the world. It is a landscape of the past, of imagination, of myth, of romantic dreams.

To many it is also a landscape where the present merges with a tourist-orientated vision of the past, an *Outlander* other land. A field of mini-cairns, a few stones high, has grown over recent years on a spot by the A87 with fine views down Loch Loyne. The purpose of these is not clear, built by passers-by perhaps to memorialise their visit, symbols of regret, remembering loved ones, a connection with the 'wild'? Amongst the anonymous miniature cairns a wooden cross, 3 feet high, 2 wide, with, attached in a plastic bag, the order of service for a cremation in the south of England dated 2015 and some handwritten details of other dead family members. On a nearby rock is painted 'Gary & Liz great friends sadly missed'. There is something unsettling about the sheer volume of these dwarf memorials, a sickly morbidity that seems to be part of the current zeitgeist.

We must, of necessity, acknowledge and remember the past, but what of the living? We must experience this landscape either by walking, climbing, cycling or swimming in it, but what is this landscape beyond the transient snapshot of the tourist or Munro-bagging passer-through? We must ask ourselves why there are no – or very few – people living here.

This landscape, for all its wild and rugged beauty, is all about power. Look closely at the wilderness of the West Highlands and consider if this is really a wilderness. Energy once provided by peat is now brought by wind, wood and rain. Above the ridge of the hill three huge turbines whirr round in the breeze. Dark green forestry plantations for timber and fuel geometrically litter this russet landscape. All around we are harvesting that energy just as surely as the farmers in the Lowlands are bringing in the crops, but the labour-intensive practices of the past have long died out, emptying this land of people.

The grey tarmac road is more organic as it flows with the contours of the land and the limitations of the fossil fuel-powered internal combustion engine. Wind, water, wood, oil and, in the churned-up tracks leading to the mobile-phone masts high above, black peat.

Down in the valley floor at Invergarry the garage is emblazoned 'Fuel Stop'. Here can be purchased petrol, diesel, bottled Calor gas –'the fuel for all seasons' – coal, logs, kindling, lump-wood charcoal and peat briquettes . . . alas stamped with the letters 'BNM' (Bord Na Móna – the Republic of Ireland's state-run peat board). A mile further down the road at the hydro-electric station water extracted from peat is used to drive the turbines.

At the end of Loch Loyne a straight, grey dam abruptly marks the transition from water to land, one piece in the huge integrated hydro-electric generating scheme which links across swathes of the Highlands. Its scope dwarfs the Caledonian Canal dug along the Great Glen 200 years ago. Water pouring down from these great heights is a power to be controlled and used.

From here on the hillside, you can trace the water on its journey from puddles to bog to burn to loch. A tunnel leads to Loch Cluanie, which then releases water via another tunnel and pressure shaft to the underground power station at Ceannacroc. Once the water has driven the turbines there, it is released into the River Moriston, which is then harnessed by the Glenmoriston power station before finally being discharged into Loch Ness.

These terrifying and mysterious surges of water can create strange wave patterns and currents on the loch; in the half-light and mist they can easily be mistaken for the neck of a water horse, the mane of a kelpie ... or the humps of a monster. But if you are of a more rational bent – and Scotland is, after all, the land of great engineers – you are more likely to be in awe of this multiple use of the same clean, natural fuel source than the swirls of Celtic mists, myth and legend. If you were one of the 99 crofts out of 100 in the Highlands in 1945 that had no electricity, you would have welcomed it. Though the very process has assisted in the disappearance of those very crofts.

Compare the journey of the water off the Lewis moor to the Atlantic. It is out to sea that we are now looking for the next generation of renewable and natural energy sources. Scotland has been at the forefront of off-shore wave and wind turbine development (much to the anger of Donald Trump, who like many second generation Scottish exiles has – or rather had – a somewhat romantic view of his mother's homeland: golf and wonderful scenery). The same debates raged when the hydro schemes went in – fears that the production of low-cost electricity would bring heavy industry to the Highlands, destroying the tourist industry. But, in fact, the fears were groundless and it was the electricity that went to the heavy industry then powering Lowland Scotland rather than the other way around.

Wind farms certainly make a mark on the landscape, but so does every other human activity in a society as developed and complex as Scotland's. It is surely impossible for anyone to naively believe that even in this 'wilderness' anything that humans can control is not controlled by them already? Look again.

We often ask who owns this land, but who owns the lochs of Scotland?

On my Bartholomew's map are listed 'Loch Garry (North of Scotland Hydro-Electric Board); Loch Loyne (North of Scotland Hydro-Electric Board), Loch Quoich (North of Scotland Hydro-Electric Board). These glacial waters can be fished for the large ferox trout – at a price.

The lochside is scarred like a First World War battlefield by cleared timber, the ground a series of trenches, the trees felled by foresters working round the clock, the incessant noise and buzz of the chainsaw, the night-time bathed in unnatural floodlight. Often it is a lone sub-contractor, as isolated as the people who drive the peat-cutting machine to the north of here on the Flows of Caithness. Once peatlands, they are snake-skinned with angular plantations of Sitka spruce and fir, planted purely for commercial reasons. On the lochside other trees grow, many imported by plant collectors serving the Victorian and Edwardian industrialists who, having made their money exploiting the poor of industrial Glasgow, not to mention those of rural Asia, Africa and the Caribbean, built for themselves grand houses and castles in the baronial manner and turned crofting communities off their land. These people became servants and ghillies, replaced on their ancestral lands by sheep, stags and grouse. Do not let yourself be deceived by the 'wild' moorland of purple heather stretching above the Forestry Commission treeline: this is controlled land, fenced, burnt, managed to the *n*th degree in the service of those who would claim to 'own' it.

The ribbon of road that twists its way through contours is each unnatural mile painted with luminous white lines, each S-bend signalled in advance by a red triangular metal warning, every chicane striped in black and white arrows as unnatural up here as a zebra; the edges of the road are marked by snow poles. Thousands of tons of rocks were blasted and embanked to support the road's passage over the hills, thousands more drainage ditches and tunnels constructed to allow the water to run off or under it, destroying many a bog. A blot on the landscape perhaps, but if you go into labour at Dunvegan on Skye you would still want the ambulance to get you to hospital at Raigmore in Inverness, 130 miles away, using a modern road rather than having to rely on the auspices of the famous fairies and their holistic traditional flag.

Every community makes its compromises with energy provision, and the peatlands are no different; people and nature are tolerated to flourish where they will, but at a price. For some this is difficult to accept.

There was recently a proposal to build a wind turbine in my village. A neighbour who was happily printing off coloured fliers against it was scunnered to answer when asked where he expected the electricity used for their production to be generated. When it comes to fuel, we are all guilty of NIMBY-ism, Not In My Back Yard. The end of coal as an industry in Scotland, whatever the political and human cost, has led to an improved environment here; as I have journeyed between Glasgow and Edinburgh in the course of my life, the bings have become less black, more green, as plant life has colonised them, and the extraction of all their component shales, grits, dross and sands has flattened them. But I still buy and burn coal, and I know that somewhere else either open-cast or deep-seam mining is still blotting someone's landscape.

We can be selective in what we see, too.

On the moors of Lewis, some will see only scars made by the cutting of peat banks and drive off fuming, but be completely blind to the Scottish Fuels oil storage and pumping facility right in the centre of Stornoway. 'Hazard', 'Danger', 'Flammable Liquid', 'Safety helmets must be worn at all times', 'Carcinogen present . . .' warn the signs at the gate.

I recall feeling the same myself when, on a beautifully snowy January day, slogging the 4,000 feet up to the top of Ben More above Crianlarich, the weather cleared as we summited and a huge panorama opened up before us, revealing the splendour of the Highlands for a good 50 miles, only to gaze south-eastwards, beyond the Highlands into the Lowlands, beyond the Carse of Stirling, beyond Blairdrummond and Flanders Moss, to where the River Forth opened out into the North Sea, and there we saw the Grangemouth petrochemical plant. I was probably still fuming as we drove home.

There are no easy solutions. Aberdeen is currently going through a recession because the worldwide price of oil has dropped from $130 per barrel to $30 in the past few years. Disused oil platforms are now moored in all the firths up and down the east coast of Scotland. Those that cannot be reused in the future are being scrapped. It was one such – the Transocean Winner – that broke free from its tug in an early August gale in 2016 and ran aground at Dalmore on the west coast of Lewis, fortunately without causing an environmental disaster. My son and I went to view it from the neighbouring headland, appropriately called Cnoc na Moin (Hill of the Peat).

Renewables are not immune from the economic downturn; even the wind-turbine business is slowing down, and with only a few prototype wave-energy generators being tested the work at my brother-in-law's Highland fabricating yard at Nigg on the Black Isle is drying up. Other recycling fuel sources are being explored. Does biomass, with its claims of carbon neutrality, deliver in reality or is it a pollutant like other fuels but wrapped in a smokescreen of sharp and fancy accounting? With fracked gas now being processed, stored and distributed at Grangemouth on the Forth after being imported from America in a fleet of specially built tankers which form a virtual pipeline across the Atlantic, whither the moratorium on fracking in Scotland?

Whatever our good intentions, for the present at least all our fuel options come at a price and that applies as much to the people of the peatlands as it does to everyone else.

There is a field of massive wind turbines on the remnant of Cobinshaw Moss commercial peat bog in West Lothian, just outside Edinburgh. I met Hew, who farms at neighbouring Crosswoodhill. He reckons the guaranteed subsidised price to the producer is £95 per megawatt/hour against a market price of £40 – the economics of the insane. And what happens when the wind does not blow? From the control room in Pitlochry, the National Grid, like the crofter, has to select the mix of energy supplies to meet the day's demands. If the turbines are not turning, where next to draw the necessary electricity that we as consumers demand?

As I look out over the peat bog and wind turbines from Hew's farm, the landscape reveals the remnants of another method of historical energy production – huge, pink pyramidal stacks or bings left over from the shale oil industry of the nineteenth century. Rocks were dug up and crushed, and the trapped oil extracted on a massive scale. This land has been mined, quarried and dug out for generations for its fuels and minerals; those less concerned with land reinstatement have left it pitted and scarred. Now enterprising people like Hew are filling these holes with the waste from our own throw-away society, capping it and drawing off the gases created by its decomposition as fuel, and recycling, shredding wood for animal bedding and selling screened topsoil for landscaping and compost for agriculture. As traditional farming became more difficult and agricultural prices dropped, they explored other ways to generate profit to sustain their family businesses. Their holiday cottages are gold-standard Green Tourism rated for environmental protection and sustainability.

In the 1990s, Hew and his wife Geraldine ran a commercial peat-cutting business as part of their farm. 'The bog had always been used to graze sheep but had been too wet, and the areas of deep peat were too dangerous to put cattle on,' Hew tells me. Using a tractor-mounted blade, which excavated the peat in long strips like plasticine, they adapted farm machinery to reduce handling and accelerate drying, the two key factors in making any commercial peat operation profitable. Extraction helped the land dry out as each year went by, so making harvesting easier.

'To promote drying, the sausage-like peats would be moved by machine two to three times before mechanical harvesting into trailers and stacking up to 4 metres high. During the winter it was screened and bagged using potato machinery,' Hew explains.

What was particularly innovative about the business was that they developed peat as a fuel for barbecues. Geraldine takes up the story: 'Barbecue fuel was further dried mechanically down to less than 10 per cent moisture in a grain-drier and broken into lumps 20–40mm in size.'

Peat is pretty wet, so I ask if the cost of drying it did not outweigh the profit made.

'Only the very black, dense peat with little water content was suitable for barbecue use, so it was quite cost effective to dry it,' Geraldine replies.

'I've never heard of peat being used for barbecues,' I say. 'How did you sell it?'

'Oh, I think we had a very good marketing plan. Once the bags were filled and stitched, they looked very attractive, with their Peatman logo, designed for us by a friend of a friend. Most of our sales were fairly local. Early on we did a few runs ourselves locally, delivering door to door. Though this was time-consuming, as always in the initial stages it was the best way to get product feedback from the horse's mouth.'

There was a great deal of interest in Hew and Geraldine's product, but the weather and outside factors led to the enterprise's eventual demise. Had it not been for wet summers, where the peat just lay in sausages on the ground, never drying out, and slim profits which spiralled into losses, along with the sudden flooding of cheap charcoal into the UK, their barbecue peat may have had a future.

The bog on the farm today does not bear the scars of their activities; indeed, without being aware that peat had been extracted it would be difficult to detect. Hew says that this is because he always cut with the blade at an angle so, unlike on the Lewis moor, the weight of the peat above the slit almost immediately closed the gap where peat had been extracted, leaving none of the crevassing I see on Lewis.

Today at Crosswoodhill there is still evidence of classic bog formation, where the trees cluster in copses in the places where the minerals gather and so provide nutrients for soil-based growth, an oasis in the desert.

Geraldine says that when she came to the farm at first she envisaged herself spending time sitting in the shade of these trees reading; a lifetime later, she confesses it never happened.

★   ★   ★

In Caithness, 250 miles to the north, the landscape has similar elements: trees (though plantations of spruce greedily drink up the moisture of the bog), peat (still being cut commercially), huge wind farms and quarries, where the very rock has been dug and mined. This Devonian rock has, for generations, provided building material for walls, chambered cairns, brochs and houses. Its flagstone structure allows it to be split in slabs, ideal for building. Beside the peat bogs generations of birds have now built their homes of bog grass and heather in amongst the stones of ruined farmhouses. There will be no more human nesting in these sad and broken chimneys, the stonework 'press' or cupboard – not changed in design or usefulness since Skara Brae 5,000 years ago – will hold no more crowdie, linen or Bible. The last flames that warmed these cold ruins were not from peat in its fireplace but from the rampaging flames of the factor's torch on the roof, fired by the ceaseless and incendiary wind, as he burned the inhabitants out of their home, driving them to the alien coast or over the oceans to make way for sheep.

The peat from the bog is cut in pellets like those formerly at Cross-woodhill but on a massive scale – huge stacks are formed 20 feet high, 60 feet long, then netted over with green mesh for drying, before the peat is shipped south to the maltings at Roseisle. There its reek permeates the barley that makes many a famous whisky. In contrast to the multi-million-pound whisky industry, there is also a thriving domestic market for peat, where you can claim 20 pence off the price of a bag if you reuse your previous peat sack.

At the very northernmost outpost of the Scottish mainland all at first appears bleak and empty; the brochs that the inhabitants built as lookout posts and defences against the marauding Vikings are now but ruins. But look closer and it is a landscape full of energy.

Out on the seas off the Caithness coast wave-power schemes are being developed. On the cliffs where the power of the waves eat into the Devonian flagstone stands the broch at Dunrath ('the fort on the mound'), which has mutated over the years into other defence establishments – Dounreay airforce base in the Second World War and the Vulcan Naval Reactor Test Establishment during the Cold War. The steel sphere of Dounreay nuclear reactor lies like a massive golf ball on the links of this vast flat land. The 'fast breeder' and 'submarine' reactors are now being decommissioned and their pulsing legacy will take many years to 'clean up'.

South of the coast the interchange of water, land and atoms remains transient. On the lunar landscape of the Flow Country, the flux tower is watching for what the human eye cannot detect. Its mission is scientific – to observe and record the areas of massive blanket bogland (at 1,500 square miles, the largest in Europe) and gauge the peat's ability to capture and hold carbon, the atoms that are warming and damaging our struggling planet. This lunar rover lookalike offers hope of an escape to a new world where the lessons of human folly have been learnt and the planet is healed.

At Forsinard the Royal Society for the Protection of Birds (RSPB) has recently built an observation tower, a twenty-first-century broch. The Flows Lookout Tower stands ever watchful through the long northern summer days and long winter nights. Across the galaxies, light years bring tiny glimmerings from other stars and the merry dancing of the electrically charged particles of the Northern Lights; fleeing from our sun, they enter the atmosphere to collide with earth atoms – oxygen and nitrogen – in a spectacular show that would embarrass a Las Vegas casino.

Beneath these intergalactic skies there are few places to hide. The disappearing ozone layer depleted partly by our fossil fuel burning is allowing the world to heat up. A drive to reduce carbon emissions in recent years has led to the development internationally of 'cleaner' power sources and in 2017 we had our first day for 100 years without coal-generated electricity in Britain. Issues still persist, however.

Diesel, which powers the 4x4s which conservation bodies use in this half-land/half-water landscape and – even more so – fuels the trains which bring visitors to the Forsinard nature reserve, has dire polluting consequences, while there are fears about its carcinogenic effects on humans and on nature (it affects bees' ability to trace the scent of some flowers). It is a challenge for those organisations which on one hand preach anti-nuclear power, anti-peat-cutting and of carbon salvation on the moorlands, yet on the other are as addicted to fossil fuels as the rest of us.

Under the Aurora turn huge wind turbines propelled by a constant earthly rather than solar wind. Clumped together molecularly on the peat jelly petri dish of the Flow Country moors, these 'green energy' providers come with their own environmental costs. Near Plover Hill, two plastic kite-shaped kites tethered to bending poles dance in the wind to deter real birds from approaching the area and its deadly whirring blades. The first windmill to generate electricity was designed by James Blyth in 1887. Born in the tiny Angus village of Marykirk, like so many we have seen in this book, he moved from country to city, becoming a Professor of Natural Philosophy in Glasgow. Retaining contact with his roots, he designed a cloth-sailed wind turbine that provided power for his summer house back in Marykirk and thus it became the first house in the world to be powered by wind-generated electricity.

The people in this almost empty landscape had no choice about leaving; they were 'cleared', driven to the coast and over the water by greedy and uncaring landowners. Many exiles hoped to strike it rich in the New World – in the gold mines of the Klondyke, many laboured with spade for little or no reward whilst back home others, briefly, flooded back into this emptying land during the Kildonan gold rush of the summer of 1869. A temporary settlement of huts, Baile an Or (Town of Gold), sprang up, but soon it became clear that any of the small amounts of gold found were hardly going to be worth the effort, and interest dwindled. You can, under strict control, still pan for gold there today: passing the site, I saw a large man in a belted dressing gown striding out purposefully from a miniature caravan, a tent with stove pipe scenting the air with the sweet peat reek in the evening air.

The 'black gold' of North Sea oil has not only fuelled an economic boom in the north-east of Scotland for the past forty years but its effects have also spread far into the peatlands, offering employment, as we have seen, to many. Now the downturn in oil that has driven people back to the traditional cutting of peat has seen other ancient technologies revisited: Aberdeen Community Energy have installed a 25-tonne, 5-metre-wide Archimedes screw supplying electricity to Donside Hydro and providing power for 130 homes, and selling into the National Grid. Every firth on the east coast is now full of parked-up rigs, and it is wind turbines that are beginning to dot the east coast horizon, much to the annoyance of the RSPB, which fears for seabirds being struck by the blades and which itself is viewed by some in the Caithness community as yet another in a long line of outside landowners imposing their will on these northern peatlands with little regard for the wishes of the locals.

Defeat, Resettlement, New Lands

Look at a map of Scotland and trace the line across the narrow Central Belt, with the Firth of the River Clyde and Loch Lomond in the west and the Firth of the River Forth in the east. In between is very little land. At the eastern end of that line lies the Carse of Stirling, low-lying, next to the river, and in the centre of the vast watery mass are Blairdrummond and Flanders mosses. To the north, the geological Highland Fault Line literally separates the country, a land of mountain and flood. To the south are the Fords of Frew, an important crossing point in the Central Belt to and from the Highlands over the River Forth that has played a vital role in Scotland's history. Lying south of the great morass of Flanders Moss, the Fords' significance has been forgotten by the draining of these marshlands for agriculture over the past 250 years, but for centuries they were a vital bridging point between cultures. When the Picts invaded the lands of their southern rivals, the Strathclyde Britons, they most likely crossed into their territory via the Fords, making for the fort at Dumbarton, but as the Welsh *Chronicle of the Princes* relates: '750 years was the age of Christ when there was a battle between the Britons and the Picts, in the field of Maesydawc. And the Britons slew Talorcan, king of the Picts.'

In this borderland between north and south, east and west, land and water, Henry Home, Lord Kames, saw the potential of both the waterlogged moss and the poor, starving, dispossessed Highlanders. Advocate, judge, philosopher, agricultural reformer and founding member of the Philosophical Society of Edinburgh, Kames was a patron to such rising stars of the younger generation as economist Adam Smith, philosopher David Hume and biographer James Boswell. He sat as a judge on the panel which found that Joseph Knight – an African-born slave sold to a Scot in Jamaica and brought by him back to Scotland – should be freed, as Scots law did not recognise the concept that a man could be a slave. He was also on the Board of Trustees for Encouraging the Fisheries, Arts and Manufactures of Scotland and, crucially to our story, he helped to manage those estates forfeited by Jacobite landowners after the '45 uprising.

This desire to make Scottish society better, his passion for agricultural improvement and his concern for the victims of high politics, like the clanspeople of forfeited Jacobites, coincided with his wife, Agatha, inheriting her family estate at Blair Drummond (Gaelic: *blar*, moss) in 1766, and led to his plans to offer large tracts of Blairdrummond Moss for 'improvement'. Kames wrote in his *Essays Upon Several Subjects Concerning British Antiquities* in the aftermath of the '45 that he thought the Jacobites entirely wrong in their belief that Scots society owed their loyalty to a king (or clan chief); rather it was land grants given as a reward for loyalty that built a society:

> Upon succeeding to a pretty opulent fortune well stocked with people . . . I clearly discovered the true meaning of the term proprietor or landholder, not a man to whose arbitrary will so much good land, so many fine trees, and such number of people are subjected, but a man to whose management these particulars are entrusted by Providence, and who is bound to answer for that trust. It is his duty especially to study the good of his people and to do all in his power to make them industrious, and consequently virtuous, and consequently happy.

His plan was a simple one. By becoming direct tenants of Kames the displaced Highlanders were treated with fairness and respect. He intended to allow the workers plots of Blairdrummond Moss, 1,500 acres of which lay on his estate. They were each given thirty-eight acres rent-free for eighteen years. This would allow them time to clear the moss and establish productive agricultural land from which they could make a living. He did not leave them to their own devices but had devised a plan which involved cutting parallel trenches or 'goats' across the moss from the higher River Teith in the north to the lower River Forth in the south. The tenants then dug out the peat and, given its high water content, it floated down the flowing trenches. Later Kames' son had an engineer build an 18-foot-high 'Persian wheel' at Mill of Tor, which controlled the water, speeding up the flow where necessary, with different properties being allocated a day or time when the water would flow through their ditches. The outline of a similar scheme can still be traced from the man-made lochan/reservoir in the middle of Flanders Moss. To begin his scheme, Kames had his Highland tenants dig down to the alluvial clay a series of roads across the Moss which were flanked by 12-foot walls of peat bog. On their own patch they dug out from these roads their shelters, like peat caves, which then dried out and which they propped up with wooden staves to make their homes. Kirk Lane, Rossburn Lane, Wood Lane, Westwood Lane, Napier's Lane, Drip Moss, Robertson's Lane and Sommer's Lane: today, these roads still exist and, passing down them, the looping line of telephone wires and poles give you an idea of how high the surface of the bog was originally. The vista westwards along Rossburn Lane is stunning, an utterly flat plain and in the distance – framed by these telegraph poles – the massive bulk of Ben Venue and the Highland Fault Line straight ahead.

The Lowland air must have been full of the Highland 'peat reek', as the excavated peat was burned for fuel, the remainder sent down the 'goat'. Some of the plots were more waterlogged than others and some farmers suffered dreadfully in the wet conditions. Even last winter the section at the corner of Rossburn Lane and Robertson's Lane was again being dug and drainage pipes added. Others used the peat that they dug out from the moss to build houses from the turfs, earning them the name of 'moss lairds' from the disparaging locals.

Although coming from only 20 miles northwards at Balquhidder, the cultural as well as geographical divide between the clanspeople and the inhabitants of Carse was exposed. These poor, hard-working Highlanders were ostracised by the local community. In the eyes of the Lowland farmers – rooted to the town and market, cultivation and European trade (Flanders Moss derived its name from the area's links with the Low Countries and Hanseatic League of northern Germany), the British state and empire – these Highlanders were the progeny of blackmailers and rustlers. To them, Rob Roy MacGregor was not a heroic warrior but a rogue; they were rebel Jacobite savages, happy to overthrow the status quo from wild unmapped glens and mists, with a different culture and world view. It is a prejudice that is still expressed today, on a November afternoon on Flanders Moss nature reserve: a Lowland couple told me quite vituperatively that Highlanders 'were given some of the moss to clear and they invited all their mates down and they cleared the lot! They blocked the River Forth with the peat they dug up and so prevented it from being a seaport.' Apart from the historical inaccuracy of this – Stirling harbour remained in use till after the Second World War and only fell out of use because of excessively high harbour dues being demanded by its (Lowland) owners – what struck me was the warmth of their disdain for Highlanders even into the twenty-first century.

By recognising that the Highland way of life was dead and that an economy stretching back to the earliest farmers based on transhumance and cattle as the chief unit of currency was rapidly being overrun by the sheep, which were replacing humans in their ancestral glens, Kames was to offer these Highlanders entry into mainstream European civic society, even if it was at the very bottom – and he did so on what were better terms than they had previously enjoyed, even in their clan lands. The majority of landlords, including clan chiefs, were happy to let 'tacksmen' (usually family members of the chief) run and sub-let their properties and farms, as long as they received their rent. The tacksmen often made huge profits from sub-letting at far higher rents than they were paying the chief.

Kames' scheme was by no means perfect, because the incomers were Gaelic speakers – the people were at first denied both education for their children at the local schools and the comfort of religion from the English-speaking parish church. It was only after the intervention of the Society for the Propagation of Christian Knowledge that Kames' son, who had inherited the estate, paid for a Gaelic-speaking teacher in 1795 and there-after children received an education and spiritual succour in their own language. It is probably a reflection on how bad public health was in general at the time that the mortality rate among the moss laird children was no worse than the average, despite the very damp conditions in which they lived. It was even claimed that the medicinal properties of the peat were beneficial.

Peat had always been cut on the moss, and the cutting for fuel was localised to near settlements. But as well as a domestic landscape it is a landscape of immense national historical value. Quivers and tremors from down the ages can still be felt on its semi-solid surface. From the moss looking east can be seen Stirling Castle, perched majestically atop its rock, the restored Renaissance hall glowing in the sunlight against the looming Ochil Hills; on the Abbey Craig, the towering monument to Sir William Wallace; and to the south the Bannock Burn, site of King Robert the Bruce's victory over the invading English in 1314, when their heavy cavalry was lured into the boggy ground and massacred – the bog as defensive weapon! The link between domestic peat-cutting on the Carse and these great patriotic historical events and people can be seen in one of the many pieces of legislation Bruce passed to re-secure the Scottish nation after the Wars of Independence. In a charter issued in 1317 reconfirming the rights of the people of Stirling, it is written:

> And that our said burgesses have been in the times fore said in the full enjoyment of digging peats in the peat moss of Skeoch, paying for each [?] one penny yearly only. Wherefore, we will and grant to our burgesses, and by this present charter for ever confirm to them that they and their successors may have, hold and possess [the said right of digging peats] in all [time to come], wherefore we command and firmly charge our sheriff of Stirling, and his baillies to keep, maintain, and defend our fore said burgesses and their successors in the said liberty.

Before Kames' experiment, attempts had been made to cultivate crops after cutting the peat: the heather burnt to release fertilisers, then oxen shod with wooden planks ploughing (or men with 'breast ploughs', where the moss was too wet to hold an ox), breaking up the peat, and clay from the base of the moss spread to a depth of about 6 inches to provide earth that would support a crop. Today these fields are still farmed. As you journey along the roads first cut out of the moss 250 years ago, houses are still being built. People are still starting new lives here.

From these moss-dug lanes, Blair Drummond House is still visible, and it is still lived in. In its grounds, animals from around the world have been resettled in a safari park. In one of his short stories James Robertson writes of working there as a student. During the very long, hot, dry summer of 1976 a pit was dug so that an elephant with sore, cracking skin could have a mud bath.

★   ★   ★

As night fell on Flanders Moss one evening my footsteps were firm where a month previously they had squelched; tadpoles had disappeared from the frogspawn-filled puddles and the only visible creature a black-ribbed slug making its own moist trail, wetly silvering the boardwalk in the moonlight over the elephant skin cracks of the bog's dry surface.

Dryness is the enemy of the bog. It could be said that this land has never been drier. It tells its story in geography, history and people – three strands woven together like the ropes and strings made from the bog reeds and mosses – through its place names.

Clay laid down by glaciers at the end of the Ice Age lies beneath much of Scotland's peat, but here it has a different source. From the north edge of the Carse, you look down on this narrow plain and it is easy to imagine a time, not so very long ago, when the North Sea flooded this land. Standing on the dune-like moraine deposits at Sandhills, fields of crops mimic the liquid movements of that time – the waves of barley, wheat ears and oats swaying and crashing on gusts of wind like water. This is a land from which the sea has retreated, but where it has left its mark. The entrances of the fields are churned to patterned ridges by tractor tyres, but unlike other Scottish fields this mud is grey and cracked hard. Here is a farm – Claylands – describing the alluvial deposits left by the retreating sea millennia ago and in its sticky clutches can still be found seashells and, occasionally, the bones of whales. The River Forth, at the south of the Carse, twists its way narrowly towards the sea, eel-like, creatures at home in this half-land/half-water environment. The coastline retreating before it can be traced by history: the city of Stirling ceasing to be a sea port; Pow of Airth no longer able to float King James IV's magnificent Renaissance Scottish fleet and its flagship the *Great Michael*, but still they build warships on the river upstream at Rosyth by the engineering prowess of the Forth bridges, which block further egress by tankers with their refined North Sea oil, or fracked gas from across the Atlantic for Grangemouth.

Turn the map 90 degrees, stand back and you see the North Sea as a river basin fed by the Forth, Tweed, Thames, Glomma, Weser, Elbe, Meuse and Rhine. Reclaimed land fringes this sea, and back on the Carse it is easy to understand why the half-land/half-water of the bog to the west is called Flanders Moss – that farm there is called East Poldar, this one Ditch End. Red terracotta pantiles from the Netherlands roof many houses, Fleming is a common surname. The wetness of the land is all in the place names: Lochfield, Fordhead, Causewayend, Pow (a marshy pool of water), Moss-side of Boquhapple, East Moss-side, West Moss-side, Poldar Moss (polder: low-lying land reclaimed from the sea), Muir Cottage, Drip Moss, West Drip, Mosslaird, a touch of the neoclassical at Nyadd Farm . . .

Climb out of the plain through Bog Myrtle Wood, Low Wood, Ashentree, Birkenwood and Heathery Hill Wood. Out of the flood plain are built big houses like Blair Drummond, the village of Thornhill, the school and kirk of Kincardine-in-Menteith, the cemetery where the dead lie – dryly – in their graves. Head north and west over a land rippled with the geological waves of the Highland Fault Line and within a matter of a few short miles you arrive at Callander, where the Lowlands give way to a different geography, history, people and language. On a map you can read the topography of this place, but also a sad history unfolds as 'shielings' and 'homesteads' give way to 'sheep dip' and 'grouse butt'. Eas Uilleam (William's waterfall), Bothan na Plaighe (the plague bothy), the cairn on Cnoc Dubh (Black Hill) memorialise unknown lives, the chambered cairn on Tom Dubh ancient ones.

When St Angus first brought Christianity here around 850 CE it is said that he recognised that it was a special place, a 'thin place', where earth and heaven were close. Tradition has it that local minister and seventh son Robert Kirk (1644–92), who translated the Bible from English to Gaelic and wrote *The Secret Commonwealth* about the fairy folk, was himself transported to their underground world and never seen again.

Back on the cleared moss 1,000 feet below it is not known who in ages past travelled to meet the otherworld through the peaty waters of the bog, but in the dry, functional, earthy Lowland cemeteries edging the Carse lie the ancestors of these ancient Highland cairn builders: McLarens and McGregors from Balquhidder; Stewarts and Fergusons from Callander; McFarlanes from Lochearnhead. One of the effects of the change from Highland to Lowland culture is the recording of birth, marriage and death in writing as opposed to the oral, bardic tradition of the Pict and Gael. In these written records you can trace the family histories as the moss is cleared and turf houses become stone, small plots and strips of cleared moss start to merge as neighbours buy out or families inherit. Censuses reveal how possession of land altered – or not. On Drip Moss number 8, William McFarlane, aged fifty-two in 1814, had inherited the plot from his father, Thomas, whom Kames had settled there in 1774 from Lochearnhead. The neighbouring plot had been divided in two by the original possessor and sold to William McFarlane's son, also Thomas, and to Alexander Ferguson, who had moved to the moss as recently as 1806.

By 1814 there were 764 people living on the moss. They had cleared enough land to support themselves, 264 cows, 166 horses, 375 hens, 30 pigs, 168 cats and 8 dogs.

In the cemetery at Kincardine-in-Menteith the surnames of Kames' first tenants are interwoven down the centuries: here lie the MacFarlanes of Kirk Lane, who married Grahams of Sommer's Lane, who married McLarens of Rossburn Lane, whose gravestones dated 1914 bring us into the twentieth century; Robertsons of Wood Lane buried their kin here in 1817 and married McCalls, who were still laying to rest their family in 2005. Many did not make it to adulthood. The 1814 Drip Moss census reveals that while many families did not suffer losses, many infants and young were carried off by consumption, croup, fever and hives in addition to the diseases of the poor – rickets, whooping cough and smallpox. Occasionally a small tragedy is described in a few words: 'John, aged 1½, killed by a cart'; 'Donald, 14, killed by a horse'; 'Elizabeth, 16, drowned'.

The family of Janet McBeath moved to the moss in Lord Kames' time. She was born in Callander on 12 May 1765. It is possible that as a girl she could have seen a visitor to Blair Drummond, who, like his host and her own family, was trying to create a better life for her generation from the morass of an old and uncaring land. Here is a letter of thanks the visitor wrote to his hostess:

To Agatha Drummond   London, Jan. 11. 1772

Madam,
I have lately received, in exceeding good Order, the valuable Present you have honoured me with . . . Please to accept my thankful Acknowledgments for the very great Favour, and for the abundant Civilities and Kindnesses receiv'd by me and my Friend during our pleasant Residence under your hospitable Roof at Blair Drummond. My best Respects to Lord Kames and Mr. Drummond. With sincerest Esteem and Regard,
I have the honour to be Madam,
Your much obliged and obedient humble Servant
Benjamin Franklin

In the 1841 census Janet is described as a 'Polder Moss Settler' and at the age of eighty-six she was living on the farm at West Poldar, where she died aged ninety-six in 1860. In the dry cemetery of Kincardine-in-Menteith a headstone dated 2005 bears the name McBeath, a Justice of the Peace. Here also lies the grave of that other judge, Lord Kames. The inscription reads:

a writer celebrated for his literary excellence / in a variety of subjects / law, criticism, morality and agriculture . . . ardent in the pursuit of knowledge / in industry and application indefatigable / distinguished by public spirit / love for his country / and zealot for promoting / every useful and laudable undertaking / The friend and protector of genius / even in the humblest spheres of life … faithful … candid and just / in friendship steady affectionate and sincere.

It was the concerted and determined approach in the age of agricultural revolution by this Enlightenment thinker which, as stated in his epitaph, was able to build friendships with some of the great men of history and yet never forget the lives of the ordinary people.

In a letter to Kames in 1769, Franklin wrote:

I am glad to find you are turning your Thoughts to political Subjects, and particularly to those of Money, Taxes, Manufactures, and Commerce. The World is yet much in the dark on these important Points; and many mischievous Mistakes are continually made in the Management of them. Most of our Acts of Parliament for regulating them, are, in my Opinion, little better than political Blunders, owing to Ignorance of the Science, or to the Designs of crafty Men, who mislead the Legislature, proposing something under the specious Appearance of Public Good, while the real Aim is, to sacrifice that to their own private Interest.

Kames was not one of these 'crafty men' and his reasons for draining the moss were not based on 'Ignorance of the Science' but on public good. Sheer will power and hard labour were combined with science and technology to make the venture a success.

By the time his son died, large areas of the peat had been cleared and a brickworks set up to fire the clay dug from under the moss to provide building materials for permanent housing. More than once I have been caught scrabbling among the ruins of old farm buildings on Kirk Lane or Rossburn Lane, examining brickwork as new 'moss lairds' move in and rebuild their large modern commuter belt houses on what was once quagmire. These buildings, the productive cleared land and the passage of time brought normality to the lives of these next generations of moss dwellers, integrating them – at various levels – into prosperous, mainstream agricultural Lowland and European society, just as Kames had foreseen.

★   ★   ★

It is one thing to cut the moss to provide heat to keep you warm on a small scale that is sustainable if the people are few in number and the bog large. It is another to cut the peat and destroy the moss to grow crops to feed you. The need certainly is great, but the result unrecoverable. Kames applied thought and science to his scheme, his aim being to feed the forfeited clanspeople, but also for them to produce a surplus of food to sell, for profit, and to turn unproductive land on his estate into a source of income, a project very much in keeping with the ideas of his friend Adam Smith in his famous book on economics, *The Wealth of Nations*. Today, Blairdrummond Moss remains farmland – last year's crops included broad beans, barley, oats and wheat, and cattle grazed its grassy pastures; it will not be moss again.

Others began clearing and draining bogs and mosses to make more agricultural land to feed an expanding population and to profit from the venture. Countless treatises and books were delivered on the best method of doing this, some, as at Blairdrummond, advocated stripping all the moss right down to the clay underneath, while others advised a partial draining and ploughing the peat with manures and fertilisers.

One such book was *The Natural and Agricultural History of Peat-Moss or Turf-Bog* by Andrew Steele. First published in 1798, he describes the wet, mossy soil and the 400 acres of deep peat bog of a farm he bought 1,000 feet up in the Pentland Hills, just south of Edinburgh. It is a farm we have visited – Hew and Geraldine's at Crosswoodhill:

a part of that great tract of bog called Cobinshaw Bog (that is, the Bog of the Herd's Wood), consisting of several thousand acres ... It must have been all once a wood, as is evident from the number of trees, principally birch, found in the moss. In its original state this bog was not worth one penny per acre, and what I have of it, in my farm, was, in the purchase I made, considered as worth nothing.

Steele describes the plants and geography of his farm and the methods he intends to employ in turning it into productive agricultural land, referencing the experiences of correspondents from around Scotland and writings from across contemporary Europe. An updated edition of 1826 included a sequel which gave the results of the moss 'reclamation' techniques that he advocated in the first edition and 'improved' by 'draining, manuring, planting trees, inclosing, building, &c ... so I have upwards of five per cent of profit on my outlay; and the value of the land is yearly increasing'. Cobinshaw Bog has 'in many places, now, merely by its drainage, acquired a surface of pasture grass'.

Steele gives some fascinating details: 'In digging a drain through this moss, my servants found, at the depth of about 4 feet from the surface, a number of ancient Roman silver medals, in great preservation.' Among the coins, appropriately enough, some bearing the image of Ceres, the Goddess of Agriculture, with stalks of corn in one hand and flaming torch in the other.

He ends:

To conclude this account ... I may with truth say, that I found it a bleak, wet, and gloomy heath, about seven miles around, without shelter, and without inclosure. It now presents to the eye, in summer, fields of excellent pasture, abounding in cattle of superior description, and diversified with thriving plantations.

Littoral: An Age of Warriors

On a glorious autumn afternoon in the south-west Highlands, we are walking with friends Nic and David across Dalrigh, 'Field of the King'. In 1306 the small remnants of King Robert the Bruce's army, reeling from defeat outside Perth and fleeing westwards, were confronted here by the powerful Clan MacDougall and defeated again. Bruce had treacherously murdered their kinsman and here they took their bloody revenge, leaving him a mere smattering of an army.

Standing looking across the boggy site of the battle, David says to me, 'Imagine standing here knowing that the guys over there are going to be running at you with swords, knives and axes, and trying to stab and slash you to death.'

Accounts of the battle report it was extremely bloody, with Bruce's horses as well as troops slaughtered by MacDougall's axemen. I think of the blade of the *tairsgeir* slicing easily through the fleshy peat. Yet within two years a transformation of fortunes saw the MacDougalls themselves slaughtered, and in 1314 on another boggy battlefield at Bannockburn Bruce won Scotland's greatest ever victory against the invading English.

Around such warriors myth is woven. Along from the battlefield at Dalrigh the boggy 'Lochan of the Lost Sword' is claimed to hold within

Opposite: *Lewis Chesspiece: Berserker.*

its peaty waters a sword he cast away in his flight, the lady of the lochan keeping it safe until it is needed again. Such appropriation of myth and association with Arthurian legend may seem fanciful in such a Highland setting, yet in the medieval cross-border propaganda war all was fair game – Geoffrey of Monmouth claimed England's heritage was more ancient, being founded by Brutus of Troy; Walter Bower's *Chronicle* countered that Scota, daughter of a more ancient Egyptian Pharaoh, was our nation's progenitor. To the south of Dalrigh, beyond the towering mountain peak of Cruach Ardrain that dominates this glen, lies, at the head of Loch Lomond, Clach nam Breatainn, 'the stone of the Britons', which is said to mark the end of the ancient Kingdom of Strathclyde, which stretched from North Wales to here; the town of Camelon, midway between Glasgow and Edinburgh (a city whose hill is called Arthur's Seat), is said to be the site of Camelot. Such are the intricate ploys and stratagems on the chessboard of high politics and power.

Through Dalrigh runs the A82. It is a natural crossroads – in every direction the names of famous Scottish warriors echo down the centuries. To the east is Rob Roy MacGregor country, to the south Stirling and the monument to freedom fighter Sir William Wallace, to the west Oban and the twelfth-century Gaelic-Norse King of the Isles, Somerled, and to the north the vast tract of peat, the Rannoch Moor and the Highland Massif. Over that moor a famous (if fictional) journey by another warrior, Alan Breck Stewart. Immortalised in writing by both Sir Walter Scott and Robert Louis Stevenson, Stewart was a real Highlander who started out fighting for the government forces in the 1745 uprising but then changed sides and fought for Bonnie Prince Charlie's Jacobites. I still thrill when reading Stevenson's novel *Kidnapped*, when he and his young companion, David Balfour, are hunted through this Highland landscape by Redcoat soldiers to the point of exhaustion:

Accordingly, I lay down to sleep; a little peaty earth had drifted in between the top of the two rocks, and some bracken grew there, to be a bed to me; the last thing I heard was still the crying of the eagles.

The warrior was just one of many roles necessary for the people of the peatlands. In the times between waiting for the spring planted crops to grow and the May cut peats to dry the men transformed themselves into the soldiers of their clan chief for a (hopefully) brief period of service. If they returned, they would resume their lives as before, wounds permitting.

At times it is hard to distinguish between warrior and pawn. The Jacobite risings changed that way of life for good; not only were the clan chiefs robbed of their power or assimilated into the upper echelons of the British state, but service as a soldier became formalised, professionalised, regimented.

Winter

THE AGE OF IRON

Agriculture and Industry

The use, destruction, plundering, desecration, rape – choose your word – of the natural resources of the earth for profit has been part of the story of capitalism for centuries. Peat bogs have not been exempt from this worldwide process. At Blairdrummond Moss we have seen the start of modern, scientific technology being applied to the extraction of peat. As the industrialisation of Scotland in the early years of the nineteenth century progressed at a terrific pace, it threw up many opportunities for extraction of natural resources for profit. Peat may have been a by-product of agricultural expansion, but it was also exploited for its own market value as a fuel, as a preservative and as a source of chemicals.

Peat, coal, iron, cotton, jute, wool, chemicals, milling, forestry, tanning and leather-working, brewing and, of course, whisky distilling – these all literally developed from cottage industries, which were small and localised concerns, to mass production, serving the fast-growing cities across Scotland, Britain, Europe and empire in the matter of a few decades.

Victorians were nothing if not innovative and, learning from people and creatures discovered preserved in bogs, attempts were made to use peat for curing leather; others tried to use its preservative and antiseptic qualities to store food for long sea voyages across the ever-expanding

Part title illustration: *Aurora over the Flow Country.*

British empire. Linen-makers exploited the acid waters of Bankhead Moss, Fife. Flax grown in surrounding fields was steeped or 'retted' in the pits they dug – known as 'lint holes' (these can be discerned on the surface of this raised bog still, which lies behind a tiny village called Peat Inn, now home to a Michelin-starred restaurant of the same name).

All over Scotland peat was exploited for profit. Often the funds to invest in money-making schemes at home had been made abroad in the newly conquered lands of empire. James Matheson had made so much money in the Far East trading opium, silk, cotton and tea with his company, Jardine Matheson, that he was able to buy the whole island of Lewis for half a million pounds in 1844. With his interest in new technology, Matheson was intrigued by a local amateur scientist who was attempting to distil peat. He helped Henry Caunter set up the Lewis Chemical Works, the aim of which was to extract resalable compounds from peat along the same lines as those being extracted from coal – paraffin for use in lamps, tarry bitumen to lubricate axles, as an anti-fouling agent on ships' hulls, and hydrogen gases, which were fed back into the factory to help heat the kilns. He helped the people of the island adapt to the new age with investment in the chemical works, but at the same time he expelled them from their homes if they were in his way and sent them far over the sea to Canada, never to return. If he wanted you out, out you would go. On tree-less Lewis his persecution went as far as having his agent remove from the roof of a croft timber beams that had been washed up on the beach because he owned the shoreline and the law said the flotsam belonged to him.

The story of the Lewis Chemical Works is a microcosm of industrial history. Despite teething troubles – which included an explosion, gassing some workers, poisoning the salmon in the nearby river and sending foul gases blowing over Stornoway – the factory ran until 1875. Eventually, the project failed due to mismanagement and fraud.

In an echo of Kames' scheme, canals were dug to float cut peat to the factory, while 3 miles of tram tracks were constructed to transport it. From New Pitsligo in Aberdeenshire to Gardrum Moss peat farm at Slamannan (*Mannan*, the Celtic bog god) in the Central Belt, the same was happening, with narrow gauge railways being cut from the landscape to transport the peat off excavated moors and mosses. So what of the people who lived by peat?

Digging for a wage is universally at the lowest end of the economic spectrum, the line between paid work and punishment often indiscernible. Despite technological advances brought about by the Industrial Revolution raw muscle power still drove the Victorian economy. These were often the only jobs available for the poor, the dispossessed and the starving.

In the 1841 Scotland census we see George Anderson, sixty, Mains of Craig, Dumfriesshire, is a farmer but also a 'peat dealer', and in Hightown Craigs, twenty-five-year-old Margaret Gillespie is a 'peat fitter'. In 1871 in nearby Cairn of Craigs, Janet Gillespie is a 'peat carter's wife'. Twenty years later, far to the north in Lerwick, Shetland, the Coutts and Fraser families are 'peat manufacturers' in Burn's Lane and Fox Lane respectively. In West Lothian, Thomas Handasyde is a 'peat moss and fire clay merchant' – his job describing the clearing of a moss and the clay underneath. In Blantyre in the Central Belt, Johan Lang, eighteen, is a 'peat worker', whilst her stepfather is a coal miner.

Whilst cutting peat for the hearth in the traditional way continued, serious steps were taken to make peat a viable fuel to rival coal. Trials on the Glasgow and Garnkirk Railway (G&GR) just to the south of Blairdrummond were carried out to test if peat could power locomotives, but in an age and location where coal was king, peat was never a serious contender, the ever present challenges of minimising handling during extraction and cost effectively drying it proved insurmountable. Peat did pose a challenge to the G&GR, though. Sometimes the peat was destroyed not for itself but because it was in the way. To reach the markets at the heart of industrial Scotland, the railway had to cross Robroyston Moss, which required extensive peat-cutting. Tree branches were then laid on the track bed, on top of which were placed a tartan of wooden beams made up of Scotch fir and red pine. Lastly iron rails weighing 12 pounds per foot were laid on this structure.

As the cities expanded and failed to cope with the influx of people from the countryside seeking work, the wealthier citizens sought to escape back to the countryside. Suburban Glasgow expanded and along train lines the Victorians built villas for commuters. A woman from Lenzie, between Glasgow and Flanders Moss, told me that a number of these villas in her town were built on the remains of a bog and gradually sank, so that living rooms became cellars. Transitional suburbia is a place of paradox. I was chatting to someone on Flanders Moss who told me his 1960s family home still had peat-cutting rights, but they were unable to burn it as the house was now in a 'smokeless zone'.

If the name of a place did not suit the suburban property developers, they were prepared to change it. Along from the drained moss at Rossie is my local train station:

> According to tradition the monks of Lindores named Mungrey 'Our Lady's Bog'. It later became known as Our Lady's Bog (*Groome's Gazetteer* iv, 449), until 1847, when the Edinburgh, Perth and Dundee Railway Co. opened the new station and chose Ladybank as the new name. After this the old name gradually fell into disuse (Gillin and Reid 1979, 28, 44) . . . The population grew from 376 in 1861 to 1,198 in 1891.

To those trapped in the alien city and their children forced to abandon traditional rural homelands, the idea of the countryside took on a romantic gloss, that varnishing aided by the worldwide popularity of Scottish author Sir Walter Scott.

For those 'steekit frae nature' in the industrial cities, a longing for the countryside came naturally. This romantic notion and modern transport technology combined so that even by the time of the Battle of Waterloo in 1815 there were sixteen steamboats plying their trade on the Clyde. The Napoleonic Wars prevented Continental travel and instead of heading south tourists began to head north. As well as the Wordsworths, the artist J.M.W. Turner left a range of stunning watercolours and paintings made on his Scottish tours.

That there was a need in a quickly industrialising world to capture the wild spirit of the Scottish hills is evidenced in the art of Turner, who sketched peat cutters in a huge Romantic landscape. But it is important not to regard the past as some romantic halcyon idyll. Turner was a 'Classical' artist who later painted in a 'Romantic' style but who was not blinded by the realities of the rural lives he depicted in the vast Highland landscape. He painted what he saw. In his watercolour sketch *Peat Bog, Scotland c. 1808* there is no pretence that these people are living in some primitive paradise.

The critic John Ruskin saw not 'happy rural toil' but 'patient striving with hard conditions of life . . . cold, dark rain, and dangerous labour'. This was the reality of self-sufficient subsistence living in the nineteenth-century peatlands. Life is hard in this landscape, for all its beauty. But there is hope. That rainbow. Sunshine comes after the rain. These Christian people would see in it a promise from God to man, a covenant made after the flood, which has maintained many in the peat-cutting culture over the centuries.

Banished from Everywhere

The northern counties of Caithness and Sutherland were rich in peat and stone but not much else. The sandstone rock is easily quarried, breaking into gravestone-sized slabs, which when slotted vertically around the perimeter of fields give the area a cemeterial quality. The ripples of history were felt, but life for the majority in these vast, bleak lands was hard and unchanging down the centuries. This was to change dramatically and tragically at the end of the eighteenth century.

In 1785 the most powerful landowner in the region, the Countess of Sutherland, married the Marquis of Stafford, whose family had become fabulously wealthy from the construction of a canal which transported coal to the factories powering the Industrial Revolution in England. With that marriage, the whole relationship between the people of that society suddenly changed. In the matter of a few short years the Sutherlands became notorious, namely in Scotland, for infamy on the grandest of scales, summed up in one word: Clearance.

Intimidation, eviction, clearance under a smokescreen of 'improvement', the Sutherlands' policy of dispossessing their human tenants in favour of more profitable sheep was nothing short of a land grab by the rich and powerful on the poor and weak, who looked to them for paternalistic protection. Whilst the process had already begun in the Southern Uplands in the late eighteenth century, it was generally slower and more organic,

and other peat-cutting societies in northern Europe took their skills and knowledge with them to the eastern seaboard of North America – the spade used for cutting peats in America is called a *slane* after the Irish term and is similarly modelled, with an extra metal wing at right angles to the cutting blade.

Some 113 million hectares of Canada, 13 per cent of its area, is covered in peatlands, with distribution across all of the country, but particularly dense in the Hudson's Bay Area of Ontario and Manitoba. With its vast forests providing immediately available fuel, it was like a throwback to the prehistoric days in the old world.

Scots arriving in Canada found a climate different from the relatively mild, wet Oceanic weather they were used to, with the winters bitterly cold. Here are two Gaelic songs that tell of the difference.

O mo dhùthaich ('Oh My Country')

They come to us, wily and cunning
To seduce us from our homeland,
They sing the praises of Manitoba
A cold country without coal, without peat.

'S fhada leam an oidhche gheamhraidh ('The Long Winters')

The winter night seems so long to me –
long and long it seems to me,
and all I see is the empty prairies –
I can't hear a single wave beating on the shore

But there was peat, and by the second half of the nineteenth century the strong contacts with the homelands and the enterprise of the emigrants led to experiments with peat as a fuel source, along similar lines as in

Europe. Starting in Quebec in 1864, then Ontario, schemes such as fuelling locomotives of the Grand Trunk Railway with peat, the use of industrial peat-cutting machinery imported from Germany and moss litter development in New Brunswick in the 1890s mirrored developments on the other side of the Atlantic.

The sacrifice of young Canadians fighting for king and empire saw the harvesting of sphagnum moss for bandages during the First World War both in St John, New Brunswick and St John's Head in the Orkney Islands. After the war the Canadian government invested in developing peat as a fuel as the price of coal rose, but by the end of the Second World War this was uneconomical. The war had created one of those strange quirks of fate that, rather than German machines, it was German prisoners of war who were cutting Canadian peat. In Ontario the Erie Peat Company employed fifty 'Enemy Merchant Seamen' to help cut on the Wainfleet Bog.

Peat-cutting as a punishment for prisoners was not unique to the New World. One of the Nazis' first concentration camps was at Borgemoor in Lower Saxony, where the regime's socialist and communist political opponents were imprisoned and forced to cut peat. Calling themselves *moorsoldaten*, they composed a song which was later taken up by the Republican side in the Spanish Civil War.

Rudi Goguel, who wrote the tune, described its first performance:

The sixteen singers, mostly members of the Solinger workers choir, marched in holding spades over the shoulders of their green police uniforms (our prison uniforms at the time). I led the march, in blue overalls, with the handle of a broken spade for a conductor's baton. We sang and by the end of the second verse nearly all of the thousands of prisoners present gave voice to the chorus. With each verse, the chorus became more powerful and, by the end, the SS – who had turned up with their officers – were also singing, apparently because they too thought themselves 'peat bog soldiers'. When they got to 'No more the peat bog soldiers / Will march with our spades to the moor' the sixteen singers rammed their spades into the ground and marched out of the arena; leaving behind their spades, which now had, sticking out of the peat bog, become crosses.

The gulags of the Soviet Union also saw political prisoners broken by the hellish toil of digging. Forced labour was used to construct roads and railways through the vast Russian peatlands, as well as for peat extraction to fuel thermal power stations, which in the 1930s were supplying the Socialist republic with 5 per cent of its power.

★   ★   ★

Digging, toil, hard labour, punishment, death. Ominously, throughout the story of peat, has been the word 'Flanders'. Flanders Fields, the digging of trenches, mud, water, hell on earth, death. Death on an industrial scale, death of the young, death of your son, your father, your brother, your brother's friends, their classmates, their uncles, cousins, even sisters, aunts and thousands at a time, hundreds of thousands, millions of deaths. And after death more digging.

Na Mairbh san Raoin ('The Dead of the Field')

Eagerly they went across the fields of strife
Who lie there stretched in everlasting quiet;
Warm was the tender breath of love from their heart's wealth
Before death's black deluge flooded and engulfed it.
In obeisance to those who fell in the battle's heat,
Beside them quietly, silently dig a grave

> And in their battle attire bury them
> Where they fell down, death to the enemy in their cry.
> Silently lift them, who won fame for glorious deeds,
> And with fond regard lay down their heads in the rest
> Time will not end through the eternity of their course;
> Close up the dwelling, and leave the lovely daisy
> To sing their virtue in the sweet breath of wind;
> And raise a cross as a memorial over warriors gone.

There is a timeless quality to Murdo Murray's verse; the warriors could be swordsmen buried under a chamber cairn. There is a reassurance about the 'fond regard' with which they are interred. But we know that was rarely the case. Whatever the actuality of their interment, the sacrifice of that generation is memorialised in many ways across the peatlands.

On Lewis, the noticeboard of the community shop at Tolsta has four crosses, four poppies, four photographs of local boys killed in the First World War. For the past few years a very moving and effective commemoration of those from the north of the island who gave their lives has been running in conjunction with both an exhibition and a book from the Ness Historical Society. Outside each croft where a man did not return there is a large red poppy. In some there are two or three. Particularly moving are the ones in front of strips of land where there are crofts no longer, or a house lying derelict in this place where the most valuable creature is the child. Most crofts run straight off the main road, so driving through you get a real sense of how many men from this community did not return. Memories run deep, and as part of the commemoration a coach from the community is, as I write, on its way to Flanders to visit the battlefields where many relatives died. Mary and John are hoping to visit the grave of Mary's uncle; these are not ancestors or forefathers, but still close family members.

When I was with Hew and Geraldine at Crosswoodhill Farm, a

woman came to the door collecting for the Poppy Appeal. The tragic loss of life suffered during the First World War all over Scotland is reflected in this amateur yet heartfelt poem from Arbroath, written in 1915:

> At eventide I keep the tryst
> As in the days of yore.
> But, oh, my heart is breaking now
> Since thou'lt return no more.
> The moon shines down as was her wont,
> And flowers bestrew the track
> But I keep the tryst alone to-night
> Beside the Old Peat Stack.

The moor that had been harvested by women for colours for cloth was now scoured for moss to staunch the flow of their husband's, son's, father's or brother's blood.

There can be few places where the futility and tragedy of the loss of so many young lives is demonstrated as poignantly as at the Beasts of Holm just outside Stornoway harbour. It was here in the early hours of New Year's morning 1919 – that most special of festivals in that most special of years, after such darkness, such lack of sun, after four horrendous years of war – that the Royal Navy Yacht *Iolaire*, bringing home islanders from the war, ran aground on the rocks; within sight of their home, after all they and their families had been through, 174 Lewismen and seven Harrismen drowned. Such utter tragedy, such despair, such anger. As the poet Iain Crichton Smith wrote in 'Between Sea and Moor': 'The *Iolaire* sank . . . bringing home from the war 200 men to be drowned on their own doorsteps, a tragedy that breaks the mind.'

Beside the official monument to those men whose drowned bodies were washed up on the very beach beside the cemetery at Sandwick where they were buried is another memorial. While the official monument is carved and dressed, four-tiered, angular, obelisk, inscribed with a crest and psalm quotation in Gaelic and English, its square-paved surround fenced in metal, immobile and immemorial like a Celtic cross, this is a simple stack of stones from the shore, round, organic, growing, each stone showing an act of devotion and remembrance, of personal interaction, communion with the dead. It is not my intention to say one is better than the other, for both serve to honour and commemorate those who have gone before in their own way, but it does hark back to that most ancient of Scottish memorials, the cairn. The same arrangement can be seen at another commemoration to lives lost at sea at the Cunndal memorial at Eoropie in Ness, where the 'official' monument is complemented by a cairn of stones.

We have seen cairns on the moors and on the tops of the hills on the Highland Fault Line to the north of Flanders Moss. Each stone a memory, an attribute, and I have suggested that there is a strong act of remembrance in the cutting, stacking and burning of peat. An explicit link between built memorials and peat stacks can be seen in Will Maclean's design of the sculptured tribute to those returning from the war who were no longer prepared to accept the status quo of the pre-war years. It combines the permanence of a technically sound, built monument with the organic construction of a cairn, with natural stone and boulder construction, like a dry-stane dyke, blackhouse gable, peat stack, chambered cairn on a grassy earth form base.

The anger and despair of the Lewis people was channelled into improving their lives, and at Gress, to the north of Stornoway, this monument is a celebration – a celebration of the men and women who took part in the 'land raids', who refused to accept that pre-war status quo and who were not prepared to accept charity from however well-intentioned capitalist magnates, but took control of their own lives by affirmative action and set up their own crofts on unused but fertile land on the island. One hundred years on, the importance of the First World

War to rural as well as industrial Scotland can still be felt. While for my paternal grandfather in inner-city Glasgow, it meant strong trades unionism and never being prepared to accept without question the words of the bosses and ruling elite (the ones who directed the same tanks used on the battlefields of Flanders onto the red flag-wielding protestors in George Square in 1919) in the peat culture it meant no longer being prepared to accept clearance and poverty at the whim of estate owners, absentee landlords and a far-off and indifferent government.

Today there are many problems facing the peatlands. Life remains difficult. But the number one aim of the Galson Estate Trust, who have taken back ownership of Ness, is the alleviation of poverty. These issues are being addressed and solutions found by the people who live there, who still believe in that 'inalienable title' between the people and their land.

The Peatlands
that Have Gone

On my way to work each morning I cross the Howe of Fife, a fertile plain of agricultural land which lies between the twin peaks of the Lomond Hills in the south and the Ochils in the north. In stretches, the road runs on an elevated causeway a few metres above the fields which, on the right, are given over to crops, supplying fresh vegetables and salad leaves to wholesalers and supermarkets. Sometimes the fields are covered in acres of polythene sheeting to promote early and late growth; sometimes they are filled with flocks of geese or swans. On the right is Freuchie Garden Centre, where I buy bagged peat for my fire. On the left is Lathrisk Farm. It is signed with a bold bull's silhouette in black against a white background.

The source of the name is Gaelic: *lios*, meaning enclosed settlement, also an enclosure for cattle (Dwelly's), and *riasg*, meaning moor, fen marsh, peat moss (Dwelly's), or natural, wet, uncut peat (*The Bird's Road*).

That the name Lathrisk has a Gaelic origin should not come as a surprise just because a thousand years later the *Gàidhealtachd* is banished to the northern and western fringes of Scotland. From the seventh century the Gaelic Scots of Dalriada in the west started expanding into the rest of what is now Scotland and by the twelfth century, when these names start being recorded, had pushed out the Picts, who left virtually nothing of their language – though they did leave us their beautiful but undeciphered carved stones. Even the name 'Pict' is the Latin one given to them by the Roman historian Tacitus because of their tattooed and painted warriors. In a peculiarity of history, the island of Lewis, now one of the last strongholds of the Gaelic language, has Norse place names – Ness, Swainbost, Laxdale. Before proper academic research, place names could be subject to myth and fancy, so the village of Dunshalt or Dunshelt near Lathrisk was claimed to be the place where Viking invaders pulled up their longships – 'Danes halt'. In reality the origin is also Gaelic: *dun* – fort, *innis* – island on a flood plain, *ealt* or *ealta* – flock or herd of animals such as cattle or birds. The name describes the location perfectly: the four small concentric rings of ditches and earthworks at the north of the village, presumably originally palisaded, were dug in the post-Roman period; the slight elevation gives a commanding view over the plain, an island in the morass; and a place where cattle are grazed in summer and where flocks of migratory birds overwinter, as they do to this day – on more than one occasion our power has been cut by swans flying into the electricity cables that are carried across the fields on pylons. This suggests not only a shared language with the modern Gaels but also the nature of the terrain and land use being similar – bog and moor with the same type of transhumance culture of summer pasturing, borne out by the name Sheils Farm, the place of the shielings. Overlooking this boggy plain was a more impressive fort of the same period (about 500–800 CE) on the pap of East Lomond Hill, suggesting a stronger, year-round occupation. During excavation, archaeologists found a Pictish stone finely carved with an image of a young bull or steer. That the Picts had a strong association with the ancient European tradition of the bull cult is evidenced by the eight bull-incised stones found in what appears to be a temple at Burghead 150 miles to the north. There is also a strong oral tradition of bull worship and myth, especially around the *tarve* or *tarbh uisge* – the water bull, which, like the kelpie, is said to reside in lochs and moorland, but not to be as evil or fearsome as its equine counterpart.

It will breed with domestic cows, their progeny being identifiable by their smaller than normal ears. That bulls are still potent symbols of sexual power and wealth in this landscape is illustrated by the petty attempts to draw a large phallus on the Lathrisk Farm sign. In 2015, 'Blelack Investment', a Charlois bull, was bought by Lathrisk Farm at the Stirling cattle auctions (just to the east of Flanders Moss).

The bleakness of this north-west European winter's day was perfectly captured 500 years ago in this Renaissance Scots translation of Virgil's *The Aeneid*:

> The dowie dichis war all donk and wait,
> The law vallis floodit all with spait.
> The plane stretis and every hie way
> Full of fluschis, dubbis, myre and clay,
> Laggerit leys wallowit fernis schew,
> Broun muris kithit their wissinyt mossy hewe.

An approximate English translation runs:

> The dismal ditches were all dank and wet,
> The low valley flooded all with spate.
> The plain streets and very highway,
> Full of marshes, boggy pools, mire and clay,
> In mired pastures shrivelled ferns showed,
> Brown moors showed their wizened mossy hue.

★   ★   ★

On a near freezing mid-December morning like this on the side of the hill between the plain of the Howe of Fife and the top of East Lomond two men and a mechanical digger prepare a new grave in the cemetery.

A wall separates the living from the dead, over which I cannot (yet) see. There must be a deftness to manoeuvring the machine between the regular gridded headstones; best practice for the task involving respect for those already resident. It's a sensitivity not given to the bog people – unearthed and undignified, put on display for all to see. Günter Grass, ordering his own coffin, making funeral preparations on the eve of his own death, wrote of his distaste for these macabre displays:

> Over the course of time no longer ours, all would decay, the box and its contents. Only bones large and small, the ribs, and the skull might remain, unlike the bodies buried in the bog in Schleswig-Holstein, now placed on show under glass in the Schloss Gottorf Museum. Those bones turned soft; you could still see tissue, skin and knotted hair, as well as bits of clothing, relics of a ghastly prehistoric age, of scientific value, eagerly sought as fodder for bog-bodies stories, like the one of a young girl whose face was covered with a strip of cloth in punishment for some atrocity that could scarcely be imagined.

Not here the irregularity of the sandy graves at Luskentyre on the Atlantic shore of Harris, an island where there is such a lack of deep top soil that the dead from far afield are transported down the 'coffin road' and densely interred with their ancestors, and where the gravestones of centuries bash your shins as you wander through the long, overgrown grass. I think of Iceland and the geothermal hotspots which prevent burial in some places, the dead having to be buried in the earth of strangers, colder than their own. Halldòr Laxness writes of the woman from the east dying in Reykjavik who, as a last request, asks for her body to be transported back home for burial.

Back in Fife, 5,000 years ago the ancient builders of the circle of posts at Balfarg, which sits incongruously within the modern circles of the many-roundabouted new town of Glenrothes, did not bury their dead. Post holes in the centre of the enclosure suggest a raised platform which may have been used for sky burial. I think of the local birds: buzzard, sea eagle, sea gull, magpie, rook, hoodie (carrion crow), corbie (crow):

> As I was walkin' all alane
> I saw twa corbies makin' a mane,
> An yin untae the ither did say,
> 'Where shall we gang an' dine the day?'
> 'In ahint yon auld fail dyke,★
> I wot there lies a new slain knight,
> An' naebody kens that he lies there,
> But his hawk, an' his hound, an' his lady fair.

The archaeologists who unearthed this circle and the bull stone high on the hill above had their own method, rules, traditions of digging. As do the drainers on the plain below – enlightened and modern environmental 'conservationists' armed with JCB caterpillars; moss lairds on Flanders Moss armed with flaughters shaped like the ace of spades ('the dead man's card'); greenkeepers refashioning bunkers at the home of golf in nearby St Andrews; and Mary and John, their *tairsgeir* slicing down into the strata of their ancestors' peat.

Cutting the top turf, digging down, re-laying the turf to grow again – the processes of peat-cutting and burial are not so dissimilar, though their purposes are the opposite. Activities at peat bank and graveside, crematorium, raised funeral platform or sacred pool in a bog are both necessary and ritualistic, marking the passing of lives and years with all their calendaristic implications.

★fail dyke – turf wall

The next morning, with the redish-orange winter dawn sun rising low over the flat lands of the Howe (at 9 a.m. at this latitude at this time of year), I pass a funeral car. Hazard lights flashing, it is parked on the double yellow 'no parking' lines at a bend in the road in Auchtermuchty – with death different laws apply. Slowly but Charon-like the lone funeral-besuited driver is walking up the path between road and house, respectfully, narrowing the gap between the private and public grief and the time between death and interment on the hillside 2 miles away, which is not yet visible across the sea of dawn mist lying over the once-bog of the valley floor. The special rules that apply for the dead and the bereaved help transport us through this grey landscape. Like the car, they are in that transient place: though the dead have made their journey into death, the living are engaged in crossing that littoral and the customs that accompany it – the funeral car; the service of remembering and giving thanks; the interment; the tea and food afterwards; the memories; the talk of those who have also gone; reaffirmation of family bonds and friendships. Tradition is so important to helping the living carry on, as in so many aspects of our lives. Families, friends, neighbours, communities come together at such times. Physically, in activities like peat-cutting, tradition ensures that the living have heat to survive the year; internally, funeral traditions help us survive loss. Each has its own rhythms: cutting, catching, throwing, stacking, burning – denial, anger, bargaining, depression, acceptance.

Standing atop East Lomond on a clear day the panoramic view is vast: to the west the high peaks beyond Flanders Moss and the Highland Fault Line; to the north the snowy plateau of the Cairngorms and Grampians; eastwards to Bankhead Moss, ancient St Andrews, then the North Sea, over the horizon to Denmark and the bog people, Günter Grass's Danzig/Gdansk and beyond again the Baltic peat-burning lands of Sweden, Finland and Russia. From the south beyond Edinburgh and Crosswoodhill Farm, Agricola marched his invading Roman cohorts.

They came not from Italy but were peat-burning Tungrians and Batavians recruited from Flanders and the Rhine estuary who fought and vanquished the Caledonians.

In his contemporary account of the campaign the Roman historian Tacitus describes the subjugation of the conquered and forced demands for tribute and taxes which led to the clearing of forests and draining of marshes.

A few metres north of Lathrisk runs Moss burn and beyond it the Rossie Drain. This artificial waterway was dug by Captain Cheape of Rossie in the first decade of the nineteenth century to drain the land and reuse it for agriculture. Our friend Mr Steele of Crosswoodhill sites him as an example of good practice in bog reclamation:

> But the finest crop of grass I ever saw on a moss, was raised by the ingenuity of Captain CHEAPE of Rossie, Fifeshire. The moss having been formerly drained, levelled, and ploughed, and the surface peat burnt, was covered over, an inch or two in thickness, with barren sand . . .

For cutting the peat, Cheape 'makes use of the ordinary turf or flaughter spade'.

Steele describes his use of burnt peat as a fertiliser: 'the ashes of swarded peat-bogs are also made use of by Captain Cheape, for improving poor lands, and he says they have a most powerful effect'.

One of the features of this landscape now are the sprinklers in summer – fed by the Rossie Drain – rehydrating the crops on this dried land. Acres of plastic sheeting cover some of these man-made fields throughout the year, protecting and encouraging growth beyond the natural seasons of the plants to provide us with perfect-looking and unseasonal foods in the volume and quantities we demand from our supermarkets and internet providers at a price that precludes it being economic for locals to make a living picking them. Agricola-like, farmers use Romanians and Bulgarians from the fringes of western Europe as their troops. But this is not a piece of agri-desert. Whilst across this plain tractors may work unnaturally by night, using satellite GPS technology to plough perfect furrows which maximise crop yield, and all-year-round cattle in massive barns are fed a diet of carrots grown under plastic, this is also a landscape rich in wildlife. Former bog residents such as skylarks and meadow pipits are still here; barn owls flourish on field mice and voles; red squirrels thrive in the oak, alder and fir originally planted by Cheape 200 years ago; deer – which in the sixteenth century had to be brought to nearby Falkland Palace for the courts of James V and his daughter Mary, Queen of Scots, to hunt – now roam naturally across this land; blaeberries and chanterelle mushrooms grow wild the woods, brambles and red gooseberries by the roadside, and herons pluck frogs from the fields. The cottages of the nineteenth-century farm workers are now occupied by Edinburgh, Dundee and Perth commuters who hang feeders to attract wren, robin, blue tit and goldfinch, and object when wood pigeon, sparrow and starling take the food (but take secret malicious delight when a hawk swoops to feast on the small birds).

It is the 21st of December, the winter solstice. In the black night sky the everlasting plough twinkles, and over East Lomond Hill I follow the bright belt of Orion that leads to the eternal constellation of Taurus, the bull.

Death, Cremation
and Burial

The beach below the cliff is swirling in the North Atlantic wind, the ground sand disappearing into the grey eternity of the lashing ocean. Six months earlier on a sunny summer's day this beach at the northern tip of Lewis – turquoise sea, yellow sand, vast blue skies, white diving gannets – had been a paradise on earth. On the same cliffs the golden letters of the newly carved names on the gravestones glinted over the sea in the setting of the westering sun into the infinity where those who have gone before (both the living and the dead) rest. Further down the coast the stones marking the sandy graves of their early Christian ancestors at Teampull Pheadair, the early Celtic church of St Peter's, have been worn anonymous by the unrelenting passage of time and wind. Further still along the seaboard at Cnip, the Bronze Age burial sites date to about 1,500 BCE. Here there is evidence to suggest that both inhumation and cremation were used when burying the dead.

I am interested in whether peat was used to fuel pyres for these cremations. There is no modern comparison because almost exclusively in the Christian tradition in Scotland the dead were buried.

The massive influx of people – including those from the peatlands – into Glasgow, Edinburgh, Dundee and other cities in the nineteenth century gave rise to the need for large necropolises: the small parish kirkyards that surrounded these ever-expanding metropoli, and which were constantly being absorbed within the city limits, could not accommodate the number of corpses being produced by the deadly living conditions. These cities expanded in a blur of industry, slum living, lack of sanitation and disease, so vividly illustrated in Dickens' *Bleak House*.

By the post-war period in this country, as Victorian municipal cemeteries reached capacity, so crematoria were built. It was as recently as the mid-1960s that the papacy granted Roman Catholics permission to be cremated and for priests to officiate at these services.

These conditions do not apply in the peatlands; geography means that even today in the Western Isles and most of the northern and western Highlands the option of cremation is not a viable one – there is plenty of land, populations remain small and crematoria are not necessary or economically viable. There is one class of exception to this universal burial practice within historical times and that is the burning of witches and heretics at the stake – their remains could not be allowed to contaminate mainstream society's consecrated ground. Before burning, witches – like the bog people – were strangled.

Cremation evidence from pre-history shows that wood was certainly used in early Bronze Age pyres, with charcoal and even wooden coffins being found in tombs, cists and burials of cremated remains, but by the middle of the Bronze Age (around 1500–1000 BCE) it is estimated that deforestation had almost completely removed all natural sources of wood on Scotland's Atlantic fringes. Such was the cultural imperative to continue the cremation funerary tradition that alternatives had to be found, thus 'Beaker People' remains in the National Museum of Scotland show clearly that in Orkney seaweed was used as a funeral pyre fuel and, as the primary cooking and heating fuel in deforested Atlantic Scotland, peat. Modern experimental archaeology has successfully cremated animal corpses on peat-fuelled pyres.

After this date there are still some traces in the peatlands of wood being used for cremation, but given its scarcity this must indicate high status or wealth.

This has echoes in the later semi-mythical story of the conquered Orkney islanders being instructed in the year 888 CE to cut peat by Earl Einarr, with the precious wood being kept exclusively for the invading Viking ruling class.

Though Bronze Age and later some Iron Age communities used peat for human cremation it was unlike our modern practice of high temperature firing of the corpse followed by the grinding of bone to produce a uniform ash for scattering or burying. Evidence shows these ancient societies burnt human remains at various temperatures and were interested in preserving larger fragments of the deceased. Flesh leaves a gradation of colours on burnt bone if it is fired at different levels of heat and the archaeological evidence from early burial places shows that the heat levels on funeral pyres varied both between sites and even within a single cremation – blue, grey/brown bone at low temperature; white at temperatures of over 800°C. Black bone points at incomplete burning. This, of course, may have been because fuel was scarce or the high temperature of a pyre was difficult to maintain consistently due to seasonal weather or winds gusting, heating some parts of the fire to high temperatures whilst other parts did not burn so well. However, a society that could smelt bronze and iron knew a lot about controlling fire. Skills were handed down and are demonstrated in practices like the use of peat-fired kilns to extract lime from sea shells, recently recreated on Barra. Slow, smoke-dried peat similar to charcoal would have been used in forges, perhaps even to make peat-cutting irons.

Donald Maclean from the island of North Uist described in 1962 making charcoal from peat:

Coal was scarce and people used to make charcoal. First, they made a hole eight feet long, three feet wide and three feet deep in the peat land. They then filled it with two cartfuls of good peat. They set fire to the peat at each end of the hole and also in the middle. When the peat was burnt through, they covered the hole with turf, ensuring that no air could get in by filling any small vents with soft peat. After two days they would take it home in bags to the smithy. The charcoal was clean and good for the smithy work. They stopped making it about 1909/10.

So, given the traditional skills in the controlling of fire and the comparison with modern cases of ritual cremation gathered by anthropologists, it seems that the uniform and complete destruction of the corpse was not of primary concern. When it came to cremation the intention seems to have been about burning the flesh off the corpse, collecting a proportion of the deceased's bone and ritualistically burying it, usually in an earthenware vessel or tomb, sometimes individually, sometimes as part of a communal burial either in a cairn or hole. There is doubt as to what was done with the remaining fragments of bone.

The ritual surrounding these burials is of interest. We cannot be clear as to the actual process that took place but the scientific and archaeological evidence does give us an insight, which touches on some of the themes and practices we have come across previously.

First, consider transformation. At death, life, which began with birth, ends. If you regard that as a purely physical process, then the circle is completed and there is no more. If, however, you believe in a spiritual or religious world, there is more and death is a transition into another way of being after life, whatever that may look like. We see in some ancient inhumations that people are dressed in their everyday clothes with their possessions, often of high value or status, buried with them to take into an afterlife. Sometimes tombs contain what may be offerings for gods.

In Bronze Age cremations sometimes something similar is going on, but there are some startling differences. Evidence from sites suggests that the bodies of the dead were burnt in pyres near where their remains were buried, but before the remains were interred some form of rite and a selection process took place. Not all the bones of the dead were placed in the tomb or burial pot. It has been suggested that some may have been given to the living as mementos or tokens of the deceased, but we do not find, for example, jewellery with human bones in the archaeology of these peoples. The remains that are transferred to the tomb or pot have been carefully selected and arranged, layering the different elements to recreate the funeral pyre but in reverse. In some cases, especially where wood was used partly or wholly in fuelling the pyre, the manner of placing the bone fragments in the pot or tomb mirrors the layering of the funeral pyre, perhaps a way of making sense of or taking control over the uncontrollable randomness of death. We cannot know the reasons for these rituals, but it is postulated that some kind of reflection of the physical world is involved.

With this in mind I would draw attention to the process of burning a body and placing the remains in the earth and compare it with the process we have seen of fertilising a reclaimed bog with burnt peat and ash to make it productive. New life is created from former life where the peat-burnt remains of one generation are interred in a traditional burial ground or cairn of the ancestors; the link with the past and the land is reaffirmed and strengthened. In the case of invasion or change of ownership it gives authority or legitimacy to the new hierarchy. The small amounts of cremated bone and the lack of grave goods in these Bronze Age rites suggest that these are token cremations; it is the placing of the dead within these ancestral places that is important rather than the individual dead themselves. This may be true also of the sacrificial victims offered to the gods we now call the Bog People: it was the rite, not the offering, that was important. The same could be said of the pyre. If by the Iron Age deforestation was almost complete in the peatlands, it mattered not to the dead how they were cremated, but the prestige that would attach to the living if they could provide wood to burn them would be great.

★   ★   ★

'Token' cremation is mentioned above and that is a theme that I want to pick up on in discussing peat in the context of another unique form of peatland burial. We have already considered the Bog People and despite Grass's distaste for our macabre interest in bodies preserved in peat, it seems to be a long-held one. At Cladh Hallan (*cladh* – mound, *hallan* – cemetery) on the island of South Uist in the Outer Hebrides, which had been occupied since the early Bronze Age (*c*.2000 BCE), archaeologists made a surprising discovery. The red/brown bones which appeared to be from a man who died about 1600 BCE and from a woman who died about 1300 BCE had been reburied about 1120 BCE, but before that there was evidence to suggest that after death they had been preserved in a peat bog for about a year before being exhumed. Their mummified bodies had then been on display in a stone-built building, presumably for some religious purpose. It is not known why they were reburied, but the bones of at least six individuals were identified. It is the only known example of peat bogs being used deliberately to preserve bodies that would then be disinterred and put on display by the culture that buried them.

In the peatlands – as in many ancient cultures – contact between the human world and the world of the gods was believed to be possible through watery places. We have seen human sacrifice in the case of the bog people; in the Scottish Borders the well of the Roman fort at Newstead had cast into it offerings of animal bones and metal objects; and at Covesea on the Moray Firth the sea-caves were used by ancient peoples as a burial site, a tidal threshold between the living and the dead. At Blackburn Mill in East Lothian there is a blend of archaeology and a fantastic illustration of how peat bogs are formed. A large hoard of iron objects was found buried in a peat bog which, at the time of deposit (0–200 CE) was perhaps a loch. A cauldron was left resting on the loch's clay bed as an offering. Slowly the loch's waters were subsumed by sphagnum mosses, which gradually layered over it, millimetre by millimetre, so that when discovered in the nineteenth century it was under 1.5 metres of peat. Within the iron cauldron, covered by another iron cauldron, was a wide selection of iron objects: farming equipment, tools, horse and vehicle fittings, vessel fittings, scrap iron for recycling, ingots, Roman objects, fittings and fastenings. The lack of personal objects suggests that this was a community offering to appease a god or gods, and such a vast array of valuable metal, which would have been of massive practical use to such a community, suggests a very great tragedy or threat of tragedy had befallen them. It also has echoes of the cremations in the late Bronze Age and earlier Iron Age, where invaluable wood was sacrificed – in both meanings of the word – in a ritual. This votive offering ties in with a pattern in which bogs often feature in Scotland at this period, for example the six shields set on the edge in a circle in a bog at Beith in Ayrshire. This has an undoubtedly militaristic feel to it, whereas the Blackburn Mill find hints at a rural community of farmers and pastoralists which has wealth only in its everyday practical tools. These tools have a mix of local and Roman origin; the sickle, which still has its wooden handle attached, is carved into a phallus, linking the gathering of the ripe harvest with the gods of fertility, thus it was both a practical and a religious object when being used to harvest crops and the duality was reversed when offered as a sacrifice. There are faint echoes of this practice even today, with the shaft of a peat iron adorned with a polished horn. The cauldron and cauldron chain would have been suspended, as we have seen, over a hearth, possibly a central, peat-fuelled one as, within the hoard, is a peat spade blade. Its design is of a type similar to those used in Scotland in recent times and is the earliest known in the whole of Britain.

Offerings of various kinds have been found in bogs across Scotland dating from this relatively late period but also from further back, as early as 3000 BCE. Bronze axeheads from the late second and early first centuries BCE are particularly prevalent, but also swords, sometimes bent or broken, perhaps deliberately. Earlier still in the Neolithic period, before the use of metals was widespread, special offerings of stone axeheads have been found across the peatlands we have travelled – Shetland, Lewis and at Gordon Moss in the Borders. If many of today's peat bogs were created by early farmers felling trees, they would have used axes similar to these. The axehead, being such a powerful tool or weapon, took on a ritual meaning as a symbol of power, as well as being of practical use; some sacrificial axeheads (such as the highly polished one made out of jadite that was found in the Borders and had been traded across Europe from the Swiss Alps, where it was made 5,000 years ago) were never actually used as tools. Recent archaeological evidence has traced strong links between ancient Scotland and the Dutch and Austrian cultures of 'Beaker People'.

I was telling Thomas, an Austrian friend, about an east Alpine jade axe in the museum in Edinburgh as we climbed East Lomond Hill one winter's day. Casually I mentioned my interest in peat and he told me that peat-cutting was a long-established industry in his homeland; indeed, his grandfather had cut it on his farm outside Salzburg until quite recently. Pictures he subsequently emailed me showed turf-like lawn-grass and beautifully angular and neat stepped banks cut into pristine blocks of peat – very Austrian.

It was not just hard objects, proto-tools, tools, weapons or metals that were given to appease or plead with the gods, but also food. Whilst the peat bogs can preserve some foodstuffs – like the thousand-year-old butter in a wooden trough from Durness – more often we are just left with the containers the offering was placed in, but what a variety there is: fat iron cauldrons studded with circular designs; thinly beaten metal bowls; chunky, roughly fashioned clay pots; open-lipped basins; pots, cups, kegs, ladles . . .

Hugh MacKinnon from the island of Eigg recounted this story in 1964 about food apparently in the bog:

A fox and a wolf found a cask full of butter at the shore. They buried it in the sand. The fox went out several times, saying he was going to a christening. When they returned to check on the cask, it was empty. The wolf realised the fox had eaten it. On the way home the wolf saw the full moon reflected in a peat bog. The fox said it was a round of cheese and that the wolf could fish it out with his tail. He put his tail in the water and it was such a cold night that the water froze and he was trapped. The fox summoned all the animals around and they tore the wolf to pieces.

Sometimes, though, it is not an offering – foodstuff, object, animal or human – that has been found in the bog but the very god themselves. In 1880 a peat cutter excavating near the narrow straight where the road from Glencoe to Fort William is broken by the ferry crossing at Ballachulish unearthed the wooden statue of a goddess in the image of a girl or young woman. The figure was made of alder wood, with two pebbles for her eyes, and dates from about 600 BCE. It had been housed in a little alcove shrine made of woven branches set up on what was at that time a raised shingle beach. In her hands she appeared to be holding male genitalia and at the base of the statue seemed to be a niche for, perhaps, placing offerings for safe passage from land to water to land again, between lochs Leven and Linnhe. But, like a peat bog, the ancient wood needed to be kept wet and in the two weeks between being resurrected and arriving at the museum in Edinburgh the goddess had shrivelled and shrunk from a young girl to an old woman, like Ayesha in Rider Haggard's *She* or the Cailleach Bheithir, the Celtic goddess of wind and storm, who can bring winter weather to any season of the year.

Winter Journey
to Rannoch Moor

The difficulty of travel in northern Scotland necessitated in times past the support of supernatural beings like the Ballachulish goddess. In the nineteenth century, to avoid the difficult mountain pass of Glencoe and the sea lochs of the coast, the West Highland railway line made its iron way to Fort William over the Rannoch Moor. Just before New Year my friend Kristan posted a photo on Facebook of his daughter and himself on this train on their way north. It must have whetted my appetite because I woke the next morning at 5.30 and on the spur of the moment decided to head for the moor.

Gathering my things together, I was on the road by the back of six, with a vague hope that I might catch a beautiful midwinter sunrise as I got there. The route, in theory, is a simple dogleg, 45 miles north up the main artery A9 to Pitlochry, then 35 due west to Rannoch Station. This westward leg is, however, not an easy drive, especially in the dark. To navigate through this land is to edge round mountainside and loch. Moss-feeding Atlantic clouds drench the mountains, constantly filling the lochs, sometimes by small streams, other times larger burns, or occasionally rivers, which have to be bridged; often these are hump-backed and the width of only one vehicle. Depending on the size of the stream these bridges can be at the lochside or built higher on the hillside, so your route undulates up and down as well as twisting in and out – it can be like sailing on a storm-tossed boat. Originally tracks used by drovers transporting cattle to Lowland markets, today's road is built on one laid down by General Wade, who after the 1715 Jacobite rising was commissioned to open up access for government troops deep into the rebellious Highlands. With military surveyors such as William Roy accurately mapping these lochs and glens, this spelled the end of the clan system that had operated in the Highlands for time out of mind. For whilst even the mightiest of past invaders – the Romans – had been content to secure the Lowlands, block off access to and from the Highlands with forts at the mouths of the glens where they met the rich agricultural lands of the east, and sail their fleets round the coast to terrorise the natives, no previous army had been able to defeat the Highlanders in their own terrain. If, initially, these roads were actually used effectively by Jacobites in the '45 uprising against the government who had built them, ultimately the destruction of the traditional Highland way of life was achieved.

The road is now edged with suburban villas, 'cottages', Victorian farmhouses, modern Scandinavian-style holiday homes, with the occasional mansion behind hedges of rhododendrons – these imported by nineteenth-century colonial Scottish businessmen and plant hunters from the mountains of India, where the Highlanders from these very glens helped the British Empire suppress and destroy other cultures and ways of life which did not conform to an imperial vision, just like their own. There are, of course, many benefits brought by these roads, but to travel them and not to see that this is a semi-suburban society set within a minutely controlled river, loch and mountain landscape would be wrong. The rise of the middle class was a nineteenth-century phenomenon and over it reigned Queen Victoria, who visited this landscape; 'the Queen's view' down Loch Tummel then, as now, a popular tourist attraction, though not, as Victoria thought, named in honour of her but after a real Scottish queen, Isabella, wife of Robert the Bruce.

As the headlights of the car illuminate verticals of chunky, pale beech trees, with the autumn's russet leaves at their bases, I zig-zag round zed bends, suddenly slamming on the brakes as a double articulated timber lorry swings round a corner ahead and, headlights blinding me, thunders past with inches to spare. Sub-contracted tree-felling using modern saws and machinery mean that timber harvesting – like mechanised peat harvesting in the Flow Country – can be done at all hours of the day and night. Loch Tummel, Dunalastair Water (there is a fine class of Victorian steam train called *Dunalastair*), Loch Rannoch slip by. Among the douce stone villas of Kinloch Rannoch there are houses for those who service the tourist industry at the resort hotel at the head of the loch – service sector jobs make up 75 per cent of the UK economy. Further along the loch day begins to break in that gradation of light diffused by the wet, low-lying cloud that imperceptibly indicates you are approaching the West Highlands rather than the Perthshire ones. Water everywhere: loch, river, burn. Wisps of cloud, like steam, hang over spruce plantations and the tiny remnant of ancient Caledonian pine forest, the Black Wood of Rannoch. In the hydro-electric generating schemes at the lochside tunnels leading from Loch Erricht in the north and waters flowing off the Rannoch Moor to the west are dammed and fed into the Gaur turbines. Like the hidden channels and peat caverns that occur naturally under a bog, there is a man-made system of pipes and tunnels interwoven under the 'wild' Highland landscape, providing electricity for the digital age. The Gaur hydro plant has been fully automated since the 1950s. Those in search of the 'authentic' or 'natural' experience shouldn't despair, as the man working at the dam told me, 'It's a' very hi-tech. Mind, ye cannae get a mobile phone signal.'

With the waters comes the start of the wilderness. This is a different Highlands to the picturesque lochside vistas. Here the hillside is all browns, blacks and buffs, like Donald John's peat stacks. Lumps of glacial deposits are immediately noticeable, transporting you to the end of the last Ice

Age. As waters flood down, the road (in most places a single, unmarked, tarmacked strip) climbs up into low cloud and it gets murkier even as dawn breaks. Here it is bleak. The predominant colours are those of the dead: the withered straw of the *Racomitrium* grass; the brown of the heather; the grey-white of the crottal- and lichen-covered glacial granite boulders. This difference extends even to the construction of the road itself, which was built eastwards from the isolated railway at Rannoch Station.

Suddenly there is movement ahead, and across the road bound four red deer and off into the hills and mist. Only a few metres further up on the other side of the road a group a dozen strong are grazing; they do not take off but nervously check me out and carry on chewing. Only when I get out of the car to snatch a photo do they gradually take flight.

It is not the purpose of this book to go into the mechanics or politics of deer numbers in Scotland, their hunting or stalking, but on this road on which cattle were once driven to trysts in Crieff and Falkirk (and where drovers sheltered under the now broken 'Heart Stone' at the side of the road), through a moor where once cattle also grazed and people lived, I got a real thrill from seeing these beasts really close up. Big wild animals in their natural habitat, for the deer have always been here. The Homeric song/poetry of the local Gaelic bard Duncan Ban MacIntyre, published in 1768, came into my head:

> Praise over all to Ben Dorain –
> She rises beneath the radiant beams of the sun –
> In all the magnificent range of the mountains around,
> So shapely, so sheer are her slopes, there are none
> To compare; she is fair, in the light, like the flight
> Of the deer, in the hunt, across moors, on the run,
> Or under the green leafy branches of trees, in the groves
> Of the woods, where the thick grass grows,

And the curious deer, watchful and tentative,
Hesitant, sensitive: I have had all these clear, in my sight.★

The road ends at Rannoch Station: a hotel (closed for the winter), the
station tearoom (closed for the winter) and a handful of houses. It is good
to be on the outside looking in at the railway; many times have I been
on the train wishing to escape the confines of the carriage and bound,
like a deer, out into this landscape. But even though it is January and I
am at a height of 1,000 feet, it is not a snowy wilderness, not today. I
have travelled through here in blizzard, the train tiptoeing across black
bog under the snow, the drifting so bad that part of the line has to shelter
under the corrugated iron roof of a snow shed at Cruach Rock. But the
warm, wet air of the Gulf Stream which nourishes Scotland's bogs is
influencing today's weather. Fifteen miles away the ski slopes at Glencoe
are empty, yet I have walked in snow down them in early June. This
unreliability of the weather and the difficulty in crossing the moor to
the ski slopes accounts for the closed hotel and tearoom (though only
15 miles from Glencoe by foot, the impassability of the geography –
quagmire moor and high mountains – necessitates a circumnavigational
journey of 85 miles by road). There has to be physical commitment to
go onward and enter into this land.

As well as a lack of snow there is also a lack of wind. Normally sheared
by blasts, today all is still, dwarf copper birches that make up the meagre
lagg fen between station and moor hang green with abundant reindeer
moss and lichens, all verdigris frills; chaffinches, long-tailed tits, goldcrests
bejewel their ruby branches.

These woodland birds have come from the forest on the other side
of the railway and it is towards there that I head, an estate track leading

★ MacIntyre fought on the government side against the Jacobites at the Battle of Falkirk.
There is a postscript to this story, which I'll tell you about later on.

westward through the trees and out onto the moor. The first transition is over the level crossing; a sign gives sage advice for my walk: 'Stop. Look. Listen.' I heed it as around the curved viaduct over the moor growls a goods engine pulling a chain of twenty-five tank wagons from the hydro-electric powered aluminium smelter at Fort William – a throwback to Scotland's now dormant heavy metal industry in a landscape forged in the lava-fed cauldron under the peat. In a twist of history – like molten rock folding back on itself – the smelter is owned by an Indian company.

The train passes, southbound for the Mossend depot outside Glasgow through embankments and cuttings. Over the railway is intermediate country: a well-maintained estate track leads through the Barracks Forest, a fir plantation; to the left the head of Loch Laidon, above the steep slopes of Meall Liath na Doire, a big grey lump of a mountain. At the edge of the forest ice coats the surface of the boggy ground. On the perimeter of the forest some trees tilt uprooted, clutching their neighbours for support, clinging on to life whilst others have succumbed to the wind and lie being devoured by their implacable enemy, the moor. When trees growing in or near bogs get too big for their roots to anchor them in the peat, the wind – 'the muckle forester' – fells them with axen blows. But once inside the wood all is green, damp, rampantly verdant with boulders, tree trunks and ground coated with lush moss, fed by myriad mountain streams and burns that etch and erode their way down from the cloud-swathed top. So productive, so well suited to this environment are these trees that the very edges of the track which are not regularly driven over are a nursery of conifer saplings ranging in size from minute frond to 3 or 4 feet tall, all densely packed; the contrast with the surrounding moorland – if you could see it for the thick-planted trees – is stark. Climbing all the way, the forest ends after a couple of miles and below a thinning band of dark evergreens the vista westward opens up: brown, mist, bleak, grey lochs, hills invisible behind cloud. The valley floor is hummocked with glacial deposits, islands in the bog; the loch has its own islands, some mere rocks protruding from the water, others, like Eilean Lubhair (the Isle of the Yew), have trees growing on them.

The wood from a yew growing on such an island in Loch Lomond where clan chiefs are buried is used to construct a fiery cross, calling the clan to war, in Sir Walter Scott's sensational 1810 poem *The Lady of the Lake*:

> The shafts and limbs were rods of yew,
> Whose parents, in Inch-Cailliach, wave,
> Their shadows o'er Clan-Alpine's grave,
> And, answering Lomond's breezes deep,
> Soothe many a chieftain's endless sleep.

Loch Laidon reaches out greedily, invading the fringes of the land in any place it can; boundaries are fluid, what is land, what water unclear. The final transition out of the trees is marked by a 2-metre-high deer fence; under an eight-barred kissing gate paved with stones is the last dry ground before the path squelches out into the moor. This land is made up of rust-coloured grasses, brown/olive heather, black peat, green lichens, green, gold and red sphagnums punctuated infrequently by big grey boulders camouflaged by more lichens and mosses, some white like the ice that deposited the stones here. In the lee of some a straw-filled hollow for animals to shelter, behind others a straw-lined pool. White ice over dark peat makes a grey carpet that binds within it grasses living, dying and dead – green, orange and yellow – like threads on the concrete floor of a tweed weaver's shed. One patch of ice is frilled like the lichens in the birch, another a miniature Loch Garry mimicking the outline of the country – Grannie Scotland with a creel on her back. Above, Stob na Cruaiche is, on a clear day, a big pointy pile of stacked rock, but today the view is disconcerting. The romantic splendour of the clear waters of a burn cascading down through boulder, heather and peat in the foreground of your vision is picked out with lucid clarity in the Highland light, but above it all is grey, blurred emptiness, as thick impenetrable cloud hangs at 2,000 feet, obscuring the mountain's peak.

In Gaelic, the burn's name means the stream of the steer with the brindled coat, perfectly describing the colouring of this land and the cattle that once grazed it. A small but no less beautiful local is the Rannoch brindled beauty moth, native here and on Flanders Moss.

As well as wild and semi-domestic animals, this landscape has been and continues to be home to wild creatures tamed – copper spaniels, golden Labradors, black and white Scotties, border collies, buff cairn terriers, brown and black brindled deer hounds. Ancient and modern myth has larger felines stalking these glens – puma, lynx or the fabled *cait sith*, a black cat with a white spot on its chest and as big as a dog; the real Highland wild cat, *Feline silvestris grampian*, is fierce enough, as this fifteenth-century description testifies:

Thare wyld cattiss ar grate as wolffis ar
With ougly ene and tuskis fer scherpare . . .

Or more recently, in the *Fife Free Press* of 24 February 1934:

Attacked By Large Cat:
Peat-Cutter's Experience In Ochil Hills

A workman's remarkable experience while cutting peat in a bog situated on Mellock Hill, one of the Ochil Hills, is reported by Mr D.J. Taylor, of Mellock Peat Products, Kinross, the owners of the bog. The man, Mr Taylor states, was up cutting and stacking peat about 10:30 a.m. when he saw a cat running down the hill towards him. To begin with he thought it was an ordinary cat hunting rabbits, but when its size was seen – almost as big again as an ordinary cat – and its intention as it got nearer seemed to be attack, he ran to a shed for a stick. The cat attacked the man amongst the material in the shed, but he managed to beat it off with the stick. He could not carry on with his work, however, as the cat sat and watched, snarling and spitting all the time. Since quite a number of hikers and others roamed over these hills, Mr Taylor considers that due warning should be given to the public.

On Lewis we encountered the Bridge to Nowhere; here is the Stile to Nowhere, a seemingly pointless stepping stone isolated when the attached fencing was removed. Also here is a bothy, Tigh na Cruaiche, its rusted corrugated-iron roof blending perfectly with the landscape, offering basic temporary shelter for transient walkers, a ghost shadow of Highland hospitality; beside it, the sad remains of a croft, a *laroch*, gradually collapsing into the moor. Like the stile, once purposeful, now useless.

Cruaiche can mean a stack of peats. I often hear on my travels from concerned people that peat-cutting destroys the land and will do untold damage to our natural heritage for generations into the future. So I often ask myself, when in these landscapes, where are the peat banks of the people who once lived here? I simply cannot see any evidence of them. On this path leading to Glencoe, where perhaps the greatest betrayal (there are so many) in Highland history took place, it is people – who were once abundant here – that are the rarest of animals. Like their peat banks, their once glowing hearths are long sunk back into the inhospitable and implacable moor and the sounds of children laughing, bards singing, cattle lowing and the old *cailleach* crying are silenced. No lamenting *coronach* sung and island burial for the chief, rowed out on his final voyage into a westering sunset, his cleared and emigrant clan dying in the unnatural slums of industrial Glasgow or the black dusty tomb of an Appalachian coal mine. Lochaber no more.

I stop, look, listen. Out here the sound of silence is startling.

Today, there is no wind and the slow, dense, dank clouds graze the empty hillsides. A herd of ghost cattle. All is muffled. A solitary crow's 'caw' occasionally breaks the quiet – few bones to be picked over here. The moor's silence invites religious contemplation, whether at the empty bell tower on the Free Presbyterian churches of the Highlands and Islands or instigated by the occasional contemplative bells echoing out from the Buddhist temple in the depths of Eskdalemuir in the Southern Uplands. To the faithful, the dangers of life on the moor can be assuaged by religion, comforting the fearful and helping them deal with the very real threat to life from above as well as below.

'Danger of Death' says the yellow warning sign pinned to an electricity pole, with an illustration of a figure being zapped. I think of the poor young women hit by lightning stacking peat in Ecclefechan in 1811, and the dangers of being out on the moor in a storm.

Slower than the Red Sea enfolding the pursuing Egyptians but already colonised by mosses and lichens, the stump of a former electricity pole, like the trees that once grew here, is being subsumed back into the watery earth. Newer, smooth caber-like tree trunks, waterproofed with peaty bitumen, carry power lines across the moor – an electric fence to keep in the giants who, local legend has it, threw the huge boulders that dot the moor at each other in a prototype Highland games with a deadly purpose. On the other side of the pole a metal plate is branded 'Cobra' – thankfully only the shy and less dangerous adder can be found here. The real biting hazard is the midge, now thankfully dormant, though even in early January hopping insects appear when my foot scuffs an exposed bit of peat.

Aware of the early sunset I turn round and allow my boots, like boats in this watery landscape, to ferry me back, back towards the living. The lowering sun is momentarily filtered through a thinner patch of cloud and the surface of the loch is silvered by it. This light catches in the cold, wet bog an occasional remnant of Caledonian pine forest, a spider's web pearled by mist in between ossifying roots, the silvering grey wood contoured like the strata of rocks once liquid, molten fire. The flaming red-tipped *Cladonia coccifera*, scarlet-cup lichen, grows abundantly and joyously here on the exposed peat, a seductive and vivacious pointillism of primary colour in a slumbering winter. Water flowing down echoes and bubbles in partially overgrown canyons and channels it has cut deep through the land. As I near the end of the moor I drink in its beauty: sphagnums – plum, lime green, lemon yellow – out of water, in water, semi-submerged; live green heather, dead grey ling, red-brown heather wood; yellow/grey dead grasses, russet tufts of grass like fur; deer hoofprints in semi-frozen peaty moss, deer droppings; crowberry leaves poking through moss and sphagnums; unique Rannoch rush (*Scheuchzeria palustris*), bog grass; pale lichen fronds like coral growing within mounds of red sphagnum and on old fence posts; waves of grasses flattened by wind and water; hard granite rocks scoured and indented by the sheer force of melting glaciers now patched with healing lichens; vaginal openings of black water in withered reed and rush beds; bellies of moss sphagnum swollen with water retention; clumps of spiky porcupines and hedgehogs of amber and sienna grasses; desert islands of moss and heather-topped boulders marooned in a straw and copper sea of withered grasses; white cappuccino foam on top of espresso brown burn; reviving coffee drunk from a flask decorated with wildlife I have not seen here – wild duck, pheasant, butter yellow spotted trout; and a peat slab of Christmas cake the same colour and texture as the vista before me. Leaving the last of my footprints of this wet land, I pass through the deer fence gate and back into another world.

I stop, look, listen. In this green world there is sound: water burbling down and tiny birds. Rather than the squelch and suck of the bog, my footsteps crunch on the sandy gravel of the track – too noisy, they silence the birds. I move to the soft moss in the centre between the ruts of tyre tracks. Better. Too often the thud of clumping boots and self-reassuring scuffing of the strident silences the quiet, soft, beautiful tweeting of gentler creatures. The forest is densely planted – compared to the moor there is no view beyond the trees that line the path – but what is lacking in vista is made up for in variety of sound. Tiny birds flit in and out of view so quickly that it is impossible to identify them by sight, their lives lived at another pace in an otherworld parallel to ours. Even when seen they are *contre-jour*, grey silhouettes against the sky: just as my eyes descry the form, colour and textures on the moor, my ears are my guidebook amongst the trees: 'prew-prews'; 'pip-pits'; 'hop-hip-hip'; 'wree-wree-wree'; ' twip-twip, weee; twip-twip, weee'; 'see-choo, see-choo'; 'chew-chew'. A brief glimpse as a cloud of 'breeps' speed by.

Finally a chance to look: two finch-sized birds are engaged in a long dispute, 'chip, chip, chipping' at each other, flitting backward and forwards between the treetops on either side of the track. Through my binoculars I pick out the dull red heads and breasts, then the distinctive beak of the Scottish crossbill – *cam-ghob* (crooked beak) in Gaelic. On the ground, pinecones show the distinctive stripping done by their beaks, quite different from that of the red squirrels that are also scattered amongst the moss. Where there are squirrels there are predators, and in the middle of the track the distinctive 'scat' of the pine marten. Slightly heavier that the native red squirrel, grey squirrels are more prone to be caught by pine martens, as when fleeing them the very lightest of branches at the tops of trees cannot support their weight and their escape prevented.

Coming out of the forest the sound that first strikes my ear is the gentle washing of waves. Here, in the middle of the Highlands, it seems odd, as does the beautiful sandy beach lapped by them at the end of Loch Laidon. I think of the North Sea flooding in over Flanders Moss, lumpen whales as big as Meall Liath na Doire hummocking its surface. This sand, ground by glaciers, is all part of this post-Ice Age landscape.

The sound of a passenger train grinding very slowly north from Rannoch Station instantly transports me back to another time, another season – a summer journey years ago on this iron road to the isles and across a boggy moor pitted with peaty lochans whose still surfaces were plated with lily pads and dotted with ice-white blooms, silent. As my memories fade like the sound of the train, my journey nears its end. Heading back to the station, I think of my son Alexander, aged eight, cuddling his polar bear on the platform at Arisaig.

The boggy Black Lochan between loch and station is plating over with ice as the day starts to end and the temperature begins to drop. There is movement on the other side of the level crossing and I can make out a man in a blue checked jacket doing something by his white cottage. As I approach I can see he is scattering chopped carrots and parsnips on either side of the road. Then, up the embankment from the lagg fen, come deer – an antlered stag and six hinds. We all stop and look, the deer listen. In an exact copy of Landseer's iconic *The Monarch of the Glen* the stag stands and stares at me. It is a disconcerting experience: here, in the flesh, right in front of me, plays out a defining image of Scotland, one of the very few symbols which represent a distilled essence of a Scotland of the mind.

Immediately a memory from one of my childhood summer holidays comes back to me – having a sketchbook on holiday on the Isle of Arran, and after seeing in the far distance deer on the mountains for the first time, wanting to draw them just like *The Monarch*. I feel sheer delight. It is like meeting Ewan McGregor and the cast of *Trainspotting* out here on the moor. But gradually that sense of euphoria starts to subside; he's not quite as big as you'd think, more a wee laird than a monarch. And I'm not encountering him on mountainside or moor but on a run-down bit of muddy tarmac with houses, garages, telephone wires and council road signs. Whilst good for business in tearoom and hotel, this is not nature red in tooth and claw but nature semi-domesticated.

At Argaty above Blair Dummond you can – for a fee of £6 – watch red kites being fed at 1.30 p.m. every day. Seeing twenty or thirty birds of prey is a magnificent sight but hardly a natural one. For encountering wildlife in their natural habitat, compare, for example, taking the funicular railway up Cairngorm to watch and listen to the ptarmigan.

Looking back, I feel quite deflated about my encounter with the deer the road on the way here — is that why they did not run away? They thought I'd been to Tesco's for a couple of bags of pre-prepared mixed veg. Do we expect too much from nature? Sometimes we like to over-dramatise our relationship with the 'wild' — the over-equipped cyclist with all the latest hi-tech gear who is barely going 20 miles; using geo-positioning for the walk along a right of way path; the tweed-clad deer stalker crawling for hours through heather and bog for the one shot.

Some would say that many who cut the peat are playing such games. That it is a leisure pursuit for retired men and the burgeoning middle classes of the isles, west and north, which has little effect on their fuel bills. This, in some cases, is undoubtedly true; however, what distinguishes it from Munro-bagging, mountain biking or grouse shooting is that it has a cultural integrity and significance that the other activities lack.

Back in my car and heading home the daylight dwindles into twilight, the mountain Schiehallion is still hidden secretly in mist and by the time I reach the forest of Faskally it is pitch dark.

The Supernatural Moor

Coming off Rannoch Moor, I mull over a day of transitions: crossing the railway track, regulated by its timetabled journeys, to the uncertainty of each footstep on the quagmire moor; the constant chirruping birdsong of the grid-planted forest after the spectral silence of the smothering mistfall out on the bog. Heading east on my homeward journey can be seen Schiehallion (when not shrouded in mist), one of those classic pyramid-shaped mountains. The name of this iconic hill translates as 'fairy mountain of the Caledonians' and for many it has a magical quality, hard to define; certainly it is a well-loved and much-climbed mountain, lying as it does at the southern edge of the Highland massif, easily reached from the populous Central Belt. Thinking about the divide between Lowland and Highland regarding the people who came from Callander and Balquhidder to Blairdrummond and Flanders mosses, it was social, historical and cultural but also an otherworldly divide; these were special 'thin' places, where the worlds of man and spirit came close. I have touched on this theme throughout my travels. In their excellent book *Scottish Fairy Belief*, Henderson and Cowan write of these liminal places where the natural and the supernatural worlds cross and of the footprints left there. People with special knowledge could read the landscape in terms of the supernatural – as well as the natural – world. They give an example of the trial for witchcraft in

1677 of Donald McIlmichall★ who recounted being taken inside a fairy hill in Argyle. The name of the place was Dalnasheen (the field of the fairy hill – compare this supernatural name with the description in the previous chapter of Allt Riabhach na Bioraich; the fairy hill is just as much a part of the landscape as the stream and the brindled cattle). Sometimes a name recalled a historical event. My friend and his daughter who were travelling on the train over Rannoch Moor were heading to Fort Augustus at the southern end of Loch Ness, where even today local names illustrate the duality of this real and spirit world: Battery Rock from where Jacobites bombarded the government troops in the fort, and the fairy mounds of Bunoich. Originally Cille Chuimein (Church of St Columba), Fort Augustus was the location in about 465 CE of one of Scotland's most famous comings together between evangelising saint and the Loch Ness monster.

> [T]he kelpie was lying low on the river bed . . . and suddenly swam up to the surface, rushing open mouthed with a great roar towards the man as he was swimming midstream. All the bystanders, both the heathen and the brethren froze in terror, but the Blessed man looking on raised his holy hand up and made the sign of the cross . . . invoking God's name he commanded the fearsome beast saying, 'Go no further. Do not touch the man. Go back at once.' At the sound of the saint's voice the beast fled in terror . . .

If the fairy site – mountain, hill, mound, loch or bog – is a crossing place, usually it is the creatures of the otherworld who come into ours – fairies at Schiehallion or the kelpie on the Lewis moor or River Oich – but sometimes human intermediaries are necessary, most famously the Bog People, who enter the otherworld as earthly messengers. One Lewis

---

★ Donald McIlmichall was also convicted of stealing a cow, a capital offence in this culture of transhumance, grazing and droving, and sentenced to hang.

superstition is that the family of a recently deceased person distributes food to the local poor, which is then received, with interest, by the dead person: '*Gheibh thu thalle e*', 'Thou shalt get it on the other side.' Are witches and warlocks human or spiritual messengers? It is on a Scottish moor that the most famous witches of all pass messages between the two worlds to Macbeth.

Now I know, like me, what really interests you in the play is not the portrayal of a man's destruction caused by his overreaching ambition, but what fuel the three 'wyrd sisters' use to heat their cauldron – peat from the moor or wood from Great Birnam? Alexander Runciman is one of many artists to interpret the story in the etching *The Witches Show Macbeth the Apparitions* (*c.*1771), but his claw-footed cauldron is magically ablaze with a necromantic symbol circle apparently fuelling it. The hubbly-bubbling cauldron is a typical three-footed iron pot of a type which for centuries could be found over peat fires either for cooking or colouring wool with plants and lichens from the moor. These are of a different design from the Iron Age cauldrons found in bogs or which would be suspended on chains over the fire.

Witches were not the only supernatural beings met on the moor who act as guides, for good or ill, in human affairs. In the Highlands a will-o'-the-wisp in the shape of a boy or *spunkie*, peaty torch in hand, would lead travellers across a bog; sometimes it might guide you safely, but on other occasions it would lead the unwary to their boggy doom.

If ethereal fairy, witch or *spunkie* could not be relied upon to guide the lost out on the moor, the comforting, steady, milky, motherly docility of a cow could, as Mary Gillies remembers:

> The cows had to be milked and on some nights when we went for them we couldn't see where we were going because of the fog. They guided us back home themselves. We would follow the cows and they would lead us back to the shieling.

So strong was the cow's homing instinct that Mary recalls:

> One night after I'd come home from a visit I fell over one of the
> animals we had sold to Alec McFarquhar. The animal was sitting
> in the doorway. I was late in coming home and was in a hurry –
> I got the fright of my life, but I never liked talking about ghosts.

Ghost, Celtic god or fairies down under layer upon layer of decaying
sphagnum till entombed in millennia of peat; time underground – a
living death – is a feature of many traditional tales. Mortals abducted by
the fairy folk think that they have only been away for a night when it
turns out that seven years have passed. Whilst we think of time in linear
terms, the otherworld operates in circles and spirals. This can apply to
these stories, too. Perhaps the greatest recent proponent of the traveller
storyteller tradition was Duncan Williamson (1928–2007). At his funeral
the eulogy reminded listeners

> that he was just the latest teller of these tales and that when you
> told a story or sang a song, the person you learned it from was
> standing behind you, the previous teller behind her and so on. It
> was this extraordinary legacy of tales and balladry from time out
> of mind . . .

That the Scottish traveller people are in themselves otherworldly, in the
sense that they do not conform to norms of society, being semi-nomadic,
different, then the prejudice we glimpsed from Lowlanders towards the
Highlanders was often true of both peoples when encountering the
travellers. In Lewis, the crofters and travellers would meet out on the
transient moor, milk, cheese, peat exchanged for tin-smithing and mainland
goods.

During the spring when we cleaned out the shielings the tinkers used to come round. They would camp down by the river and spend most of their time at our houses. The older women would give them tea and whatever they had themselves – not one of them was allowed to leave without eating something first. One woman came over and asked for some sheaves of corn, in return she gave my sister three plates . . . My grandfather met the woman in the door just as she was leaving. When he saw the corn that she had he took it away from her because our own animals needed it, so that meant my sister had to pay for the plates in another way. They didn't really mix with us, so I didn't get to see the insides of their tents.

The travelling folk were a link to the hunter-gathering tradition, whereas the crofting culture of farming, fishing, shieling and transhumance was nearer the settled permanence of the agriculturalist and town dweller. That travellers cross in and out of day-to-day life is referenced by Robert Burns in both 'The Merry Muses' and in one of his greatest poems. If *Macbeth*'s witches are the most famous in literature, then the most famous Scot pursued by witches is surely Tam O'Shanter. In this classic poem, recited by tradition from memory annually at Burn's Suppers on the 25th of January around the world, we begin by seeing Tam and his pal, Souter Johnnie, settling down to an evening's drinking by the warm fireside in a tavern in the market town of Ayr. The working day is done and travelling pedlars have left the town's streets:

When chapmen billies leave the street,
And drouthy neibors, neibors meet,
As market days are wearing late,
An' folk begin to tak the gate;
While we sit bousing at the nappy,
And getting fou and unco happy,

We think na on the lang Scots miles,
The mosses, waters, slaps, and styles,
That lie between us and our hame,
Where sits our sulky sullen dame.
Gathering her brows like gathering storm,
Nursing her wrath to keep it warm.

Burns writes of the long, sobering journey home over the transitional ground – 'mosses, waters, slaps [steps], and styles' – between merry drunkenness and sobering reality; carefree irresponsibility and the harsh realities of working poverty; flirting with the barmaid and his abandoned, hurt and angry wife; an 'other' world and this world. It is this understanding of the human condition that reader, and listener, down the centuries and around the world recognise all too well.

Despite the danger, Tam's foolish shout of 'Weel done, Cutty-sark!' has a mischievousness about it which is associated with the Scottish supernatural. As every peat cutter should know, a spade left in the turf overnight will be used by the fairies, but only for mischievous ends.

Materials that had lain silently in the half-world of the preserving peat and then exhumed from the dark stuff of the bog could have magical properties. They could be used for more than mischievous ends. As recently as the beginning of the twentieth century. F. Marian McNeil writes about an object used by a witch that can be seen in the National Museum of Scotland in Edinburgh:

A few years ago, the present writer came across a witch's cursing bone ... which had been the property of an old woman, a reputed witch, who lived near the head of Glen Shira, in Argyll, and who died there at the beginning of the present century [in 1900]. Such was her reputation that even after her death none of the Glen people would touch any of her possessions, and it was the minister

. . . who found the bone upon the window ledge in her cottage. He took it away as a curiosity . . . in 1944 it was presented . . . to the Scottish National Museum of Antiquities.

According to tradition, when the 'witch' wanted to 'ill-wish' a neighbour, she took her cursing bone and made her way to his croft between sunset and cock-crow. She did not go into the dwelling house, however, but made for the hen-house; and seizing the hen that sat next to the rooster (his favourite), she thrawed its neck and poured its blood through the hollow bone, uttering curses the while.

The bone, which appears to be that of a deer, has been stained by age to a deep ivory. It is enclosed in a ring of dark bog oak, roughly oval in shape. This is obviously a phallic symbol, to which the 'witches' were notoriously addicted.

Note the cursing bone was on the window ledge, a transitional place in the house, which features in the story Duncan Williamson tells about a little girl who appeared to a crofter in the form of a sea otter. After changing back into a girl.

she went to the window and looked out across the bay. 'Shall I tell you what's on board that ship?' she asked. 'It's meat,' he said. 'Would you like it delivered to the shore or your house?' she asked. 'I don't want any more of your tricks!' said Ian, knowing now she was a witch.

But the girl took no notice of the man and went to the fire and took a bowl of water back to the windowsill. Then she drew a comb across the face of the water and on its surface she scattered a handful of peat. Ian tried to speak, but no sound came out of his throat; he tried to raise himself up out of his chair but the power was out of him and not a muscle could he move. Then the girl blew upon the face of the water and slowly, round and round with one finger, began to stir the glore and as she did this, Ian heard the wind begin to blow and the black clouds of a storm came over the sea . . .

Once more Ian struggled to raise himself from his chair but he was paralysed, like a dead man; he just sat there helpless as he watched the ship begin to spin and the sea was whipped into a fury of spindrift, spray and spume. He struggled to cry out. He thought his last breath had come when suddenly the door was flung open with a bang and the girl's mother stood there, panting. When she saw what was going on she rushed to the window, grabbed the basin of water and flung it on the fire. 'Enough of these games and wickedness!' she said. 'Go home to bed at once!'

Well, the girl stood there silent – then she smiled and looked at Ian, and she went to the door, where she turned and did a little curtsey, before running home. After that the storm abated and the strength gradually came back into the man. But I've heard that Ian walked like a cripple for three days after that, for that girl had witchcraft sure enough!

Catacombs – whether real passages and caverns formed under and through the peat bog, or imaginary parallel worlds under fairy mounds, or the homes of *spunkies* on the moor – are woven throughout the structure of the Atlantic peatlands. Are we in the twenty-first century now in a post-magic, post-religion world?

★　　★　　★

I began my journey through this 'otherworld' landscape in the twilight, coming off the moor past the fairy mountain of Schiehallion. By the time I reached the eastern end of the road at Faskally it was pitch dark.

This is where annually in October at the time of the autumn equinox and Hallowe'en an outdoor light show amongst the trees attracts thousands of visitors. Tickets are limited, access restricted, passes are checked and checked again. You must assemble in Pitlochry town centre and are transported to the forest by specially commissioned coaches. Drawn from bright, cosy, centrally heated homes by the magical allure of this landscape guests are transitioned into a different world, a strange blend of nature/technology and a little bit of magic. You are entertained by ethereal music, coloured lights, dry ice, automated woodland spirits and traditional storytelling in this twenty-first-century 'Enchanted Forest'. I have already booked tickets for this year.

Littoral: Snow-blind
A Narrow or Wide View
from the Tower?

Height, like being out on the open moor, gives perspective. Does being high up gazing over an expanse of bog give you even more? Yes, perhaps . . . and no. Today the snow is falling in thick flurries over Flanders Moss. It comes in the greyest, coldest waves driven by strong winds, blinding and bland-ing everything. Then it relents and a bright blue break in the clouds lets sunlight pick out the glint of molten watery pools among the snow-rounded tussocks of heather and warms the variations of blushing red on each bullfinch's breast. It highlights every individual blade of bog grass curving under the weight of snow, which then springs up as the sun's imperceptibly warm rays melt the ice-cold oppressor that has gripped them. But even in the sunlight the cloud hangs thick and low, partially obscuring the view of the foothills only a mile distant. To the west, where the peak of Ben Lomond can usually be seen, all is the utter grey of the next snowstorm as it lines itself up. Within seconds it is here, blinding me again. After the sun's brief visit the moss is not frozen but very near it, so the snow penetrates the surface of the watery pools and rests on the submerged sphagnums, turning the texture of a slushy grey/green cocktail – yet another variation of the thousands of land/water mixes on the moss. But the snow makes even the firmest of grounds as treacherous

Opposite: *Rannoch Brindled Beauty Moth (Male).*

as the surface of a bog; on my drive here, the main road was thick with it, pure white, no slush, and even at 20 mph the ABS was flickering off and on as the tyres lost their grip at the slightest turn of the wheel – tyre grips and snow chains flicker through my mind. From the muscular tyre tracks on the causeway out to the moss, I could see that only one other car had ventured here this morning, though long gone, as the snow hurried/flurried to fill in the tread.

Hundreds of crows lined the fence by the track and, like a long black wave breaking on a beach of white snow, sequentially took to the air as I drove past. The elaborate camouflage of pheasants is suddenly turned useless by the weather, wood pigeons faring better; a good day to be a mountain hare or ptarmigan. To my right the clay field that last summer was lush with broad beans is now desolate, their black stalks short pencil strokes on a blank page, but underneath the snow their decomposition is enriching the land for this coming year's crop.

Once at the car park, decisions have to be made about footwear: rugged treaded walking boots for warmth but which restrict where you can venture, or thin but higher and waterproof wellies for sploshing action in amongst the peat. Even with the thickest socks I own – thick, oily wool, hand-knitted on South Uist – my feet are cold within a minute in my wellies, so I keep moving.

Today the entrance to the moss is magical. In addition to being draped with the usual lichens, the silver birches of the lagg fen are coated in snow – silver and white, a blush of pink – forming a Gothic fairy-tale tunnel of a winter wonderland gateway, utterly enchanting. In the midst of winter the childhood delight of Narnia returns. If the supernatural silvery trees are lightened by the snow and their upper branches pearly like a delicate spider's web, then the angular framework, rigid verticals and black wood of the tower looks even more menacing – an evil lord's dark keep. I begin climbing its stairs. The horizontal steps start off wood – protected by the landings above – but, as I ascend, turn by turn they become snowier until at the top I am leaving deep footprints. Looking back at my spiralling trail, I think of twisting beanstalks and the Minotaur's labyrinth. But at the top there is no giant to grind my bones into snow-white bread, no bull/beast, only far in the distance the lone lowing of a milky, motherly moss cow.

On the viewing deck the snow lies untouched over the wooden slats, leaving its surface furrowed like a ploughed field. Out on the moss the grid pattern of the former drainage channels are clearly whiter than the squares of snow-topped heathers and grasses. The path snakes whiter still out into the wilderness of the moss proper; somewhere out there the man-made adder den of stacked branches is invisible. Looking directly down, I see layers of colour and texture: brilliant white mashed-potato snow; the vertical supports of the tower's balustrade – sodden, black, oak-mottled with lichenous green; the pale turquoise of a protective glass panel reflecting, like water, the vertical lines of the flooring planks, badger-striped beneath; cold stainless-steel clips and bolts, even their tiny edges holding snow; far below, the swirls and whorls of the snow-whitened surface of the moss. The imagination blurs and I am gazing down at the Minch over the top rail of the Hebridean ferry.

Looking directly up, I expect to see gannets but hear only geese and, far away, the dull thudding of heavy farm machinery rather than the ship's engine as it thuds on, ploughing and plunging into the next snow-capped wave.

Or am I looking down on this Flanders landscape with the Flemish eye of Bruegel, the pools in the moss the same gloomy blue-green as the frozen rivers and ponds in *The Hunters in the Snow*, these waters/lands framed by high mountains, real or imaginary? But no community labours through the months here, and the only hunters today are avian.

Twenty miles away on Black Moss, Armadale, the reservoir that was dug out of the bog in the 1860s, served each winter as a curling rink for human pond-skaters.

Curling, first mentioned as a sport in Scotland in the records of Paisley Abbey in 1541, is also depicted being played in the paintings of Pieter Bruegel the Elder, dated twenty years later. That it was Dutch peat cutters who worked at Black Moss in the nineteenth century merely continued the long tradition of the links between Scotland and the Low Countries. *Bonspiel* (literally 'good game') – the word to describe a curling match – has Flemish origins.

I pour a coffee from my flask and open the tin foil (which in its manufacture has probably passed twice over the Rannoch Moor) to enjoy the last piece of our Austrian New Year stollen cake, its icing sugar-coated surface perfectly camouflaged against the landscape below. Down on the surface of the moor all is white, but below the peat is rich brown, holding within it preserved vegetal riches – candied peel and fruits of a rich, iced Christmas cake – delicious! Not so appetising are the small snow-covered clods of dead grasses which lie like white upturned woodlice on a grey slush, spotted with rusty birch leaves like foxing on an old and crumbling book. White flecks this surface as another wave of snow blasts in from the west. A tiny nut-brown wren hops under the snow-capped roof of a clump of heather, reminding me of its compatriot back in August on the North Lewis moor hopping down 'the little bird's road' and into the shelter of a *rùdhan*, the small stack of drying peats.

The snow not only thickens and dissolves difference, it also drains colour, so that the horizon between moor and sky smudges into a blank nothingness. The driving snow forces your head down, focuses your eyes on the ground before you, and on your own footsteps. Toes numb with only thin rubber and wool sock between snow and icy water, it is these steps that reveal the colours and textures hidden by the snow – between compressed ice ridged by the grip of the boot's sole shades of brown peat striped by green and gold threads of bog grass or a lattice work of dead stalks, copper, tawny, as varied as a pheasant's plumage or a woven creel, appear. Again the flurries ease and first the trees of the lagg fen appear faintly, then in the distance the lone Scots pine towards West Moss Farm, where my son Joe and I saw a beautiful black-and-white moth on Father's Day half a year ago. As the skies briefly clear, the yellow lichens growing on the rail of the bridge over a pond dazzle with the intensity of their primary colour, the pure white boardwalk zig-zagging across the speckled moor, pulsing blue before my dazzled eyes. Looking back at my footprints, my eye catches the twin caterpillar tracks of the diggers that were at work on the moss last autumn, two distinct lines of dark slush among the white. Snow melting makes the fur trim of my tweed hat resemble the saturated green sphagnums in these grey pools. Today it is the tweed that outshines nature, a memory of what is hidden under the winter's snow and a promise of what will be in the coming spring, summer and autumn; in fact, what has been in the untold seasons this moss has experienced for time out of mind.

But not for ever. Back up in the tower for a last solitary look as the weather closes in again, blinding the view, I see the people who I have met and remember previous conversations held here: the American visitors delighted to share a place visited by one of their Founding Fathers Benjamin Franklin; the elderly lady who remembered looking down from the peak of Ben Venue at the moss fifty years ago; the worker for the non-governmental conservation organisation who looked at a tree and saw only 'bad nature' intent on drinking up the moss's valuable water; the other who said, 'If only we could restore the whole of the land between here and Stirling to pristine bog.' What of the children of the displaced Highlanders who cooked a hot meal on peat fires dug from Blairdrummond Moss, or were these poor, powerless people 'robbers', as described on the SNH website, and where do they expect the food, picked so easily from supermarket shelf, to be grown?

Why restore the bog? It was only one – very short – period of this landscape's many transitions. I would love to stand in this tower and, looking down, instead of remembering being at sea, see to the east Stirling castle rock an island after the North Sea has refilled this inlet and to the west see thousands of people sporting on the beautiful sandy beach it would recreate under the slopes of Ben Lomond. Imagine the pleasure of standing here and watching whales transport with delight before Ben Venue, their skin mottled with barnacles like lichens. Or back further and watch with awe as lava, hotter than any peat fire, bubbles up, and the rocks fold and buckle as Ben Venue and Ben Lomond are thrust up by tectonic forces as the Highland Fault Line is created.

In the future both these scenarios will probably happen again, but for the present the moss seems to be in a good place: protected by legislation, cared for by government agency, water levels managed, recognition of the past, an education programme, with monthly 'meet the ranger' events in spring and summer, working with local farmers and landowners; there is a group on Flickr sharing wildlife photographs, and there are many people who just love the place. Some come here to escape people, others come to meet and talk. Many share tips on where and what to watch, both here and in surrounding places, what's seasonal, what's rare. Some have astounding knowledge, others none. Because it is boardwalk, pram and wheelchair access is easy. Some are happy just to walk their dogs, others question how a dog running around can possibly add to a human attempt at interacting with wildlife (what of those who pick up their dog shit and fling the bag into the bog?). It does concern me that people are often the last consideration on the minds of those who are in charge of managing the natural areas on behalf of the people, and I don't mean people in the abstract, like school trips organised where sets of quotas for visits on one side are ticked off against sets of quotas on the other. There is also the quality of the experience. Bogs are dangerous, but we should be encouraging people to get their shoes off and feel the peat between their toes, both literally and metaphorically; people's experience of the moss should not be limited to the boardwalk and a boardwalk mentality. It is not unreasonable for the bog to be treated as much as possible as it has been for the preceding millennia and part of that, surely, is both the grazing of domestic as well as wild animals, which is now happening on a very small scale, but also the burning of the surface vegetation to help eradicate the spread of trees, as was always done in the past.

Back home I sketch a design for a boardwalk that goes down into the bog.

Spring Again

THE AGE OF FOLLY
OR A NEW GOLDEN AGE?

'In nova fert animus mutatas dicere formas / corpora'
('Changes of shape, new forms, are the theme
which my spirit impels me now to recite.')

– Ovid, *Metamorphoses*, Book 1, lines 1–2.
Translation David Raeburn, A New Verse Translation,
London, Penguin, 2004

This has been a book of transformations – sphagnums, peat, people, hours, seasons, millennia. As the wheel of the year turns full circle, what will the coming year bring? Will it be as Ovid postulates in his *Metamorphoses* an Age of Folly or a new Golden Age?

In our culture the moor is seen as a bleak, uninhabitable, wet, windswept place where wild emotions and evil happenings are experienced. But the peatlands are also a place of lightheartedness and joy, rich in flora and fauna.

We are in the *faoilteach*, the last two weeks of winter, first fortnight of spring: on which side will the future fall?

Strathspey Journey

Opening the hotel window, I let in the air, full of the sound of this coastal landscape, piping oystercatchers and woodland twittering, beach and beech. The peach February sunrise shines horizontally onto the vertical grey trunks of the mature trees, re-smelting the autumn's fallen leaves copper again. Angie's Lewisian ear picks out the drumming of a woodpecker – spring is on its way! Even here in the north, Nature will out. The 'Ping!' of Twitter alerts us to a new message: a cafe in Japan, land of the rising sun, where you can cuddle and feed hedgehogs with your tea, a concept as unnatural as a woodpecker on Lewis. We marvel at the folly of our fellow humans before going for a swim inside the seaside hotel and running for thirty minutes on a treadmill without going anywhere.

Breakfast is scrambled eggs and peat-smoked salmon. Our route is like a Speyside malt whisky map – Glen Moray, Longmorn, Benriach, Glenlossie, Glen Elgin. Between Mannochmore and Speyburn, Fiona and Mike are starting married life together in a house in the woods, their neighbours red squirrels and pine martens. Circling Rothes are Glenrothes, Glen Grant, Glen Spey, Caperdonich and the Macallan distilleries.

I got chatting to a man who told me a story of whisky galore. When

Part title illustration: *Tomintoul Peat Moss.*

he was a boy, it was discovered that a pipe high up in the rafters of the local distillery had twenty-four neat little holes drilled in it. Each was cunningly bunged with chewing gum and impossible to see without close inspection. Subsequently police and excise officers raided many a house in and around the Speyside village early one morning. Once things had settled down his father took him for a stroll on a moonless night. First they passed through fields of barley, then headed uphill till they came to the open moor. Either he pulled too hard or the suction was too strong, but when retrieving a bottle from the bog in which it was hidden the string slipped from round the neck and the whisky was accidentally sacrificed, reclaimed by the peaty waters from which it came.

Whilst whisky flourishes, religion is in decline, and on a Sabbath morning the old kirk on Rothes High Street is now an antiques shop. I dig among the rusty tools for old ditching spades or peat irons. Heading southwards, we cross the river beside Thomas Telford's early iron bridge, commemorated by Marshall's fine fiddle tune 'Craigellachie Brig' – a Strathspey reel whose crossings and weavings are reflections of the bridge's formal struts and the river's free-flowing dancing. 'Strathspey' is such a beautiful word, yet everywhere signs proclaim 'Spey Valley'.

Rant and reel over, then on to Aberlour. Benign just now, in spate the Spey will change from crystal clear to muddy brown and overrun the valley floor. We like to imagine – due to generations of skilful marketing – that Scotch whisky's journey is from cask to bottle, very romantic, but let's remember this is a multibillion-pound international business, those distilleries circling the towns, dominating local employment. One of our family's favourite car games when on the A9 heading north into the Highlands is to play 'Spot the McPherson's of Aberlour whisky tanker'. To give you a flavour of how big this industry is, we passed tanker number 279.

Out of the town Glenallachie, Dailuaine, Benrinnes, Knockando, Glenfarclas, Ballindalloch, Cragganmore, Glenlivet, Braeval, Tomintoul. From salmon fishing bothy and Speyside Way path, the road rises through agricultural land, snow poles incongruous beside ploughed fields but a necessary feature of this half-tame/half-wild landscape.

Between Glenlivet and Tomintoul the road climbs away from the Spey and high into the hills. Through this land flow numerous peaty burns; on our right, Josie's Well and Blairfindy Well, where the Glenlivet draws most of its water. On the pale gold-coloured moorland of mat straw grass, Monadh Buidhe – 'the yellow heath' – regrown banks hint at previous peat-cutting. Angie protests when I suggest a gentle boggy investigation, so on we drive. The day is glorious, and even up here at the edges of the Cairngorms National Park there is a feeling of spring in the crisp air – the sky clear and blue, the vistas panoramic and sharp. The hill of the eagle – Carn na h-Iolaire; to the west Strathavon, the hills then haughs of Cromdale, and into the Highland massif. The traditional song commemorating the Battle of the Haughs of Cromdale celebrates this liminal as the border between Highlands and Lowlands but it is not just a north/south one but here an east/west one too, between the mountains and the fertile coastal lands of Angus, Aberdeenshire and Moray (note that even the clothing is liminal – tartan trews being of Highland cloth but trousers of Lowland design):

As I get up bae Auchindoun
A'neath a wee bit Dufton toun
Tae the Heilands I was bound
To view the Haughs of Cromdale.

I met a man with tartan trews
an' speared at him whit was the news,
Quo he, 'The Heilan' army rues
That e'er it cam tae Cromdale.

The song tells of the 1690 battle where Jacobite-supporting Highlanders were defeated by the English, but then it turns history on its head and merges this loss with a victory of fifty years earlier under the command of the dashing Marquis of Montrose:

The Grant, MacKenzie and M'Kay,
As Montrose they did espy,
Then they fought most valiantly
Upon the Haughs of Cromdale.

The Gordons boldly did advance,
The Frasers fought with sword and lance,
The Grahams they made the heads to dance,
Upon the Haughs of Cromdale.

And the loyal Stewarts, wi Montrose,
So boldly set upon their foes,
Laid them low wi Highland blows
Laid them low on Cromdale . . .

The people of the Highlands are different from the Lowlanders, a difference that often leads to bloody and savage battle. The 'Lowlander' can be Scot, Roman, Viking or Englander, but suppression of the Highlander and their culture is always the aim. It is naive to think that this no longer goes on today, sometimes consciously, but mostly unthinkingly. We Scots are children of a fractured history, victims of bullying who can ourselves bully those even gentler than ourselves.

That this area has been inextricably linked with peat and peat-cutting is not in doubt. These are from the Ordnance Survey name book of the area, which was compiled in the mid-nineteenth century by the field agents and cartographers drawing up the OS maps. They interviewed local people of all classes to find out what the names of the features they were mapping were called.

> Feith Musach, 'An extensive moss in which the fuars of Tomintoul and the farmers of Strathavon and Glenlivet, have a right to cast peat for fuel. Meaning the filthy bog or myre.' OS 1/4/18/67
>
> Feith Geal, 'An extensive moss. common to the inhabitants of Tomintoul and district for a peat moss. the name signifies the white bog.'
>
> Tom na Moine, 'A small knowe near Lagganvoulin. The name means the peat knowe.' OS 1/4/18/46
>
> Feith Dobhrain (Faindowran), 'A small peat moss used by the inhabitants of Tomintoul for fuel. Meaning 'the otters bog', 'at the north west end of the planned village of Tomintoul'.
>
> Feith an Eich. 'The name signifies the horse's mire or bog.' OS 1/4/18/66

Just before the road drops down, the surface of the high moor turns dramatically from white to brown. Here is the last remnant of Tomintoul peat-cutting, a small commercial operation hemmed in by road and spruce plantation. But not for long – the planning permission granted by Moray Council in 2006 was only for twenty years and, given the current climate, it seems impossible that it will be renewed. (When I returned four months later all the peat-cutting machinery was for sale.) There are, rightly, strict guidelines on what is permissible – cutting is only allowed to go down to 0.5 metres of peat to enable regeneration; once cutting is complete the bog must be re-seeded with local grasses and plans put in place to stop conifers from adjoining plantations seeding on moss. By granting the permission the Council recognised that peat was a local fuel source used within 50 miles of the cutting and thus had a low carbon footprint, provided local employment and had a unique cultural heritage both domestically and within the whisky industry.

So cut peat is integral to the culture of this place, the smell of its burning a feature of this landscape as much as the tussocks of yellow grass, but will it join the otter-less bog and horse-less mire as a meaningless name on the map? Out of the 4,500 kilometre square landmass of the Cairngorms National Park, the 0.12 kilometre square area of peat-cutting is 0.0027 per cent. It is the only place within the park that this traditional form of fuel provision is practised. What, then, when it disappears? How can we have a national government agency with the duty of care called Historic Environment Scotland and let such an historic Scottish environment disappear from a national park? How can we care so much for the preservation of the flora and fauna of a place and yet allow ancient human tradition to become extinct before our eyes?

Some would suggest it is the same anti-Highland attitude I touched on earlier in another form. Some would suggest it is a class thing: influential, rich landowners happy to see more land for grouse and deer – to shoot. Once the cheap peat goes, increased fuel costs will be another factor forcing the poor out of tied rents in potentially valuable local properties – to their landlord's glee. For now, with substantial grants for peatland restoration to aid carbon capture, the land is better unworked. Others argue that with the growth of wildlife and eco-tourism, the same landowners are shifting their focus, welcoming an influx of conservationists and naturalists with a narrow agenda and lack of cultural understanding. These, combined with outdoor adventure holiday businesses, have brought into the area a middle class who only commit to this land for a middling amount of time, have little knowledge or understanding of local tradition and history, and who, by employing seasonal labour on wages that make year-round jobs and housing unsustainable for many born and brought up in the Highlands, are speedily destroying native culture. What does it say of a people who live in a land where they cannot pronounce the name of the hills, hear the voice of its waters, speak the language of the moors?

The visitor becomes the centre of this landscape, those who come to witness and experience living nature and those who come to kill. 'Nature' is reserved, or 'game' is killed. The lines are often blurred; those who abhor the shooting of pheasants imported from Asia happily agree with the eradication of American grey squirrels that are overrunning the local red squirrel population (those who would see themselves as liberal and would never consider themselves racist use the language of the far right when talking of these furry incomers). Between the two camps the reintroduced osprey, the 'success story' of the Cairngorms, can be spotted with assured regularity from the timeshares at Dalfaber in Aviemore on the feeding run between the nest and the trout fishery. 'Deer park', 'grouse butts', 'visitor centre' replace 'shieling', 'farm', 'school'. Campsites increase whilst traditional travellers' sites are reduced and fenced in with legislative 'protocols'. Reindeer and beaver are 're-introduced', as Grants, MacKenzies, M'Kays dwindle to extinction. On the moor the fat cuckoo lays eggs in more of the tiny meadow pipits' nests.

Summer Visit

The Earth belongs unto the Lord
And all that it contains
Except the Clyde and the Western Isles
For they are all MacBrayne's.

I am travelling back through the strata of journeys to the first year my
wife took me to the island where she grew up. We are on the quayside
at Ullapool, waiting to board the ferry to Lewis.

There is anticipation on the jetty as the cars and passengers line up.
Feet fidget before the gangplank; fingers drum on car steering wheels;
husbands who had assured their wives it was not necessary to book pass
anxious looks to Caledonian MacBrayne staff as they wait to see if there
will be any spaces; children, adults, dogs from all points of the compass
agitate excitedly to be on board – some pleasantly, some irritatingly.
Round the point the black and white bulk of the *Isle of Lewis* glides into
view. I thrill in seeing a large object in a small space. The ferry from
Stornoway looms over the low, white one- and two-storey houses and
shops of the town as, with spiralling watery Catherine wheels from its
side thrusters, it tip-toes delicately to a shuddering halt against the pier
and the engines stop with a clap of silence that echoes down Loch Broom.
Ropes are thrown, making first contact between island and mainland. A

yellow lion rampant roars on the red funnel as the last plumes of engine smoke are absorbed by the pure Highland air.

A klaxon warning sounds and the bow of the ship miraculously splits and starts to rise, ramps and walkways descend and the transition between water and land begins. Cars laden with bikes; holidaymakers in campervans and caravans; coaches full of tourists; natives off to visit mainland friends – all appear out of the belly of the boat.

Once they are gone it is our turn. In an incomprehensible, secret order known only to the initiated, sailors weigh up how to balance the boat and start directing the lines of waiting cars into the hold and the ritual of embarking the vehicles begins. For three hours they will be held in suspended animation, moving but not moving, transport within transport.

Safely parked we ascend from hold to top deck through layers of ship – car deck, lounge, play area, cafeteria, bar, shop – from a dark metal box lit by electric light and echoing with the idling turbines' hum to bright nature of mountain and loch and seagull.

On the ferry the tone of the engines moves from neutral and the deck starts to quiver, radios crackle, voices shout and ropes are cast off. Translucent purple jellyfish pulse at the ship's departure from its dock. The voyage begins within the green fjord tightness of the westward-facing loch. It still feels incongruous being on this huge, throbbing vessel in this 'wild' lochside landscape, yet the Clyde and its Highland sea-lochs were the very birthplace of the steamship trade and for the past 200 years they have been the connecting threads shuttling between Scotland's communities. These journeys form the weft and weave of Scotland's cultural cloth.

Onboard we are higher than the former church, now host to the Ullapool Museum, where peat iron, creel and model emigrant ships are all on display. We slip past the houses as it too recedes into history. Among the pleasure boats and wildlife cruisers a silver salmon, all muscle, nearing the end of its Atlantic crossing, breaks the surface and, arcing its glittering scales back into the grey, rids itself of more sea-lice, the power of its will and instinct driving it on a journey further than this boat will ever sail, across a thousand miles of ocean.

Mountain, water, weather: the tripartite powers constantly dominate here. Ben Mor Coigach looms over the ferry as we progress towards the sea; at the foot of its pink Torridonian sandstone slopes the mouth of the loch opens to gulp down the salt waters of the Minch. On the northern shore, one of the tiny white houses that lie like daisies on the green sward of lawn between mountain and sea is the Achiltibuie Garden. There the hydroponic method of cultivation is practised in greenhouses and poly-tunnels – minerals and nutrients needed by the plants are delivered via water rather than in soil, allowing the gardener total control over growth. Beyond is a landscape Man can do little to control – the stunning frieze of the Assynt mountains has not grown but is the remnants of a landscape once carved by the cold power of retreating ice caps. These remnants of layers of soft pink sandstone, some capped with quartzite, rise from the geologically fascinating land – Stac Pollaidh, Cul Beg, Cul More, Suilven, Canisp, Quinag – each an individual whose character changes as the boat enters the sea proper and begins to leave the land far behind.

Southwards the tattered fringes of this coastline open up views down similar mountain-sided sea-lochs to the one we've just left and on the myriad tiny islands mottled seals bask on yellow-lichen encrusted rocks, ready to slip back into the water at a second's notice. Further out between us and the north coast of the Isle of Skye, black and white seabirds bob on the ever-deepening swell – guillemots, eider ducks, black-headed gulls.

Eyes are peeled for whale and dolphin, but today it is a school of porpoises who are following in the boat's white and turquoise wake. They are out-of-the-ordinary creatures. They are big animals. The delicate scales of the salmon are as tiddlers beside them. I thrill with their energy and vivacity, but they make me feel out of my depth. Whilst the seals returned from land to sea, humans evolved into land-walking creatures, so whilst I envy these mammals their mastery of water, it is a habitat that brings me fear. Licking the salt spray off my lips, I grip the guardrail for some feeling of security on this calm, summertime crossing, the regular bars of angular man-made metal lock me in from the undulating, irrational heavings and pitchings of the ever-changing waves. Evacuation procedures, lifeboats, even on one crossing the reassurance of a practice by the coast-guard helicopter winching people off and on the boat, partly comfort. But out there amongst the porpoises I would not last a minute, salt water on my lips, in my mouth, up my nose, in my lungs – no matter a spray-whipping helicopter or fluorescent orange lifebelt.

The colour of the sea at a distance is grey, but staring down over the guardrail the water directly below is a deep dark green flecked white and whorled by the froth and spume of the bow wave. Looking up, buff-headed gannets glide marvellously by, their narrow wings and harpoon-sharp beaks making them the perfect dive bombers for fish, as one turns on its own wingtip and plunges from air into water, transformed from T-shape to V.

People, too, descend for food – fish and chips, macaroni cheese, mugs of strong tea or something stronger in the cafeteria or bar and – depending on wave and weather – stagger and roll, whether they have had a drink or not.

Back on the green-painted deck some are sitting on red plastic chairs looking back towards the Assynt mountains, now suddenly no more than pale blue lumps on the horizon; others look north, where, if you were lucky enough to find land, it would be the Faroes or Iceland, and the sea, whether you call it the Minch, the Atlantic, the Norwegian or Green-land Sea, is implacably just sea. Amongst our fellow passengers are a jarl and his whole squad of Shetland Island Vikings resplendent in burnished metal-winged helmets and armour, a twenty-first-century reimagining of Norse warriors heading for Ness, Swainbost, Laxdale, and the gala day parade in Stornoway.

On a day where sea, sky and the thin strip of land between the two are all grey, a shaft of sunlight illuminates the island capital as we approach our journey's end.

Excited, and relieved to be nearing dry land, I ask Angie how she feels returning.

'*Gu leor*,' she replies. *Happy enough.*

Ebb and Flow

I am standing among the enigmatic jumble of ancient stones on the edge of the moor above Steinacleit on the far north-west coast of the far north-west island of Lewis. Among the sheep droppings and bog cotton wool tufts, a glorious psalmody of larks and meadow pipits is rising. The stones are ancient in both geological and human terms. Whether they are burial cairn, stone circle or domestic settlement remains unclear, but what they tell us is that people have been living successfully in this narrow land between ocean, upland moor and sea for at least the last 5,000 years.

The living is hard. Humans are one of the rarer animals; over the past 100 years the population of Ness has declined from 3,000 to 1,000 people. Before me is a smattering of crofts. Long, thin strips of land run at right angles to the grey tarmac road that runs parallel to the coast, on each a one- or at most two-storeyed croft or detached house, maybe thirty of them. There is uniformity and not much difference: people live pretty much the same here. If you lived here, you would probably be able to name every person and child in a 5-mile vicinity. The Free Church is at the centre of the community. There is no shop and the school has recently closed; however, life continues – if there is one thing that this scene tells us, it is that people adapt. And not only people:

The nest of the raven
is in the hawthorn rock.
The nest of the ptarmigan
is in the rough mountain.
The nest of the blackbird
is in the withered bough.
The nest of the pigeon
is in the red crags.
The nest of the cuckoo
is in the hedge-sparrow's nest.
The nest of the lapwing
is in the hummocked-marsh.
The nest of the kite
is high on the mountain-slope.
The nest of the red-hen
is in the green-topped heather.
The nest of the curlew
is in the bubbling peat-moss.
The nest of the heron
is in the pointed tree.
The nest of the stonechat
is in the garden dyke.
The nest of the rook
is in the tree's top.

Beneath the slope, where the circular stone structure was dug out from a metre-deep covering of peat 100 years ago, is a lochan, and in the middle the remains of a fortified broch – like a stone nest – reached by a causeway. On its banks are the only trees in this vast panorama of moorland and ocean, a few wind-sculpted Scots pines. Yet here herons have built their unlikely nest. They will eat fish and crustaceans from the sea; fish, ducklings

Ferguson, Andrew, *The Ghost of War* (The History Press, 2016)

Glob, P.V., *The Bog People* (Faber and Faber, 1969)

Grass, Günter, *Of All That Ends* (Harvill Secker, 2016)

Gunn, Neil M., *The Silver Bough* (Whittles Publishing, 2003)

Gunn, Neil M., *Sun Circle* (Canongate, 2001)

Gunn, Neil M., *Whisky & Scotland* (Souvenir Press, 1998)

Haldane, A.R.B., *The Drove Roads of Scotland* (Birlinn Ltd, 2011)

Henderson, Lizanne and Cowan, Edward J., *Scottish Fairy Belief* (Tuckwell Press, 2001)

Hunter, James, *On the Other Side of Sorrow* (Birlinn Ltd, 2014)

Jefford, Andrew, *Peat Smoke and Spirit: A Portrait of Islay and Its Whiskies* (Headline, 2005)

Kesson, Jessie, *Glitter of Mica (*Black & White Publishing, 2017)

Killin, Mhairi and Watt, Hugh, *Re-Soundings* (Graphical House, 2016)

Laxness, Halldór, *The Fish Can Sing*, trans. Magnus Magnusson (Harvill Press, 2001)

MacCaig, Ewan (ed.), *The Poems of Norman MacCaig* (Polygon Books, 2005)

MacDonald, Aeneas, *Whisky* (Canongate 2006)

MacDonald, Donald, *Lewis: A History of the Island* (Gordon Wright, 1990)

MacDonald, Finlay J., *Crowdie and Cream* (Time Warner, 2003)

Mackenzie, Compton, *Whisky Galore* (Penguin Books, 1963)

MacIntyre, Duncan Ban, *In Praise of Ben Dorain* (Kettilonia, 2013)

MacQueen, John and MacQueen, Winifred, *A Choice of Scottish Verse 1470–1570* (Faber and Faber, 1972)

McLean, Allan Campbell, *Hill of the Red Fox* (Floris Books, 2015)

McNeill, F. Marian, *The Silver Bough* (Canongate, 1989)

Manley, Chris, *British Moths* (Bloomsbury, 2015)

Morrison, Alex, *Before I Forget* (Alex. Morrison, 2012)

Neat, Timothy, *The Summer Walkers* (Birlinn Ltd, 2002)

Ovid, *Metamorphoses, A New Verse Translation*, trans. David Raeburn (Penguin Books, 2004)

Pearson, Mike Parker, *From Machair to Mountains* (Oxbow Books, 2012)

Piggott, Stuart, *Scotland Before History* (Polygon Books, 1982)

Ramsay, Allan, *The Gentle Shepherd* (1725)

Ramsay, Allan, *The Tea-time Miscellany* (1733)

Reid, John, 'The Carse of Stirling in the 13th and 14th Centuries' in *Calatria*, 30 (Falkirk Local History Society, 2013)

Robertson, James, *365 Stories* (Penguin Books, 2014)

Robinson, James, *The Lewis Chessmen* (The British Museum Press, 2004)

Rotherham, Ian D., *Peat and Peat Cutting* (Shire Library, 2009)

Royle, Trevor (ed.), *In Flanders Fields* (Mainstream Publishing, 1990)

Scottish Literary Tour Company, *Land Lines* (Edinburgh University Press, 2001)

Scott, Walter, *The Lady of the Lake* (Birlinn Ltd, 2016)

Shaw, Margaret Fay, *Folksongs and Folklore of South Uist* (Birlinn Ltd, 2005)

Smith, Albert Henry, *The Writings of Benjamin Franklin*, Vol. X (Haskell House, 1970)

Smout, T.C., *Exploring Environmental History* (Edinburgh University Press, 2011)

Steele, Andrew, *The Natural and Agricultural History of Peat-Moss or Turf-Bog* (Edinburgh, 1798, new ed. 1826)

Stevenson, Robert Louis, *Kidnapped* (Canongate, 2006)

Tacitus, *Agricola and Germania*, trans. Harold Mattingly (Penguin Books, 2009)

Thomson, Derrick S. (ed.), *The Companion to Gaelic Scotland* (Blackwell Publishers, 1983)

Thompson, David, *The People of the Sea* (Canongate, 1996)

Márkus, Gilbert and Taylor, Simon, *Place Names of Fife*, Vol. 4 (Shaun Tyas, 2010)

White, Kenneth, *On Atlantic Edge* (Sandstone Press, 2006)

Williamson, Duncan, *Fireside Tales of the Traveller Children* (Birlinn Ltd, 2009)

Acknowledgements

I am very much indebted to the many people who have shared their peat stories with me in the writing of this book. One of the great pleasures of researching is listening, and whether it is being told on the boardwalk of Flanders Moss nature reserve what birds to look out for, or given a lesson on the Gaelic names for each process of cutting out on a Western Isles moor, or how water levels on Bankhead Moss, Fife, are monitored, these conversations have all informed my writing.

On Lewis I am particularly grateful to Anne Mcleod at the Ness Historical Society, North Dell, and the many people connected with the society who gave of their time to assist my researches. To the people of Ness, when I turned up at their doors asking about peat stacks or came squelching over the moor to talk about cutting techniques, I acknowledge your freely given help – and tea – with my thanks. No one was more generous than Mary and John Campbell, Skigersta, who have, in their helpfulness and kindness, been as exemplary an example of island hospitality as you could imagine. My thanks to Agnes Rennie and the team at Acair Publishing, Stornoway, for their help and stories, and to Anne Campbell, author of *Rathad an Isein/ The Bird's Road*, for her help with the nuances of moor words. All factual or grammatical errors are mine, rather than the contributors.

On the mainland my thanks go to Geraldine and Hew Hamilton, Crosswoodhill, for their help in my researches, kind hospitality and their gift of a bag of barbecue peat. I am grateful to Dr Lauren Parry of Glasgow University for sharing her knowledge of peat coring on Kirkconnel Flow, Dumfries. I am extremely grateful to her colleague David Borthwick, who has been a source of inspiration and encouragement throughout the writing of this book and who supplied me with a spare pair of his trousers on the occasion of my near successful – if unintentional – effort to turn myself into a bog body. Thanks to William Simpson, Avoch, for sharing his knowledge, experience and his good company.

To Jenny Brown, who had faith in me and my book enough to act as my agent, I am extremely grateful. To Hugh Andrew, I am honoured to be published by you –

few have contributed more to the cultural well-being of the Scottish nation than you over the past twenty-five years and I am proud to be allowed the opportunity to contribute even in this small way. For their editing skill, professionalism and patience I am full of admiration for Andrew Simmons and Deborah Warner at Birlinn.

To Angie, my wife, I pay loving tribute. Little did she know that day walking in Ness when she replied 'On you go then,' to my comment that someone should write a book about peat stacks that she would lose her husband in a fug of peat smoke for the next three years.